Modern Biology and its Human Implications

J.A.V. Butler, D.Sc., F.R.S.

Crane, Russak & Company, Inc.
New York

ISBN 0–8448–1007 X

Library of Congress Catalog Card Number 76–27619

First printed 1976

First published by Hodder and Stoughton Educational

Copyright © J. A. V. Butler 1976

2869316

Published in the United States by
Crane, Russak and Company, Inc.
347 Madison Avenue, New York, N.Y. 10017.

Printed and bound in Great Britain.

Contents

Acknowledgements

I am greatly indebted to the following who have given me photographs or other illustrations with permission to reproduce:

Dr. H.E. Huxley, (M.R.C. Laboratory of Molecular Biology, Cambridge); Dr. F.H.C. Crick, (M.R.C. Laboratory of Molecular Biology, Cambridge); Professor M.H.F. Wilkins (King's College London); Professor C.A. Thomas Jr., (Harvard Medical School); Professor M. Tomoeda, (Kanazawa University, Japan); Professor E.G. Gray (University College, London); Professor R.E. Dickerson (C.I.T., Pasadena); Doctors M.S.C. Birkbeck and E.H. Mercer, (Chester Beatty Research Institute, London); Professor W. Hayes, (Canberra, Australia); my son, W.H. Butler (St. George's Hospital, London); Professor E.J. Ambrose and Dr. D.M. Easty (Chester Beatty Research Institute, London); also to The Society of Authors for permission to include a quotation from *Everybody's Political What's What.* by G.B. Shaw (1944).

Preface

Science is able at the present time to explain the nature of living organisms and how they work, to an ever increasing extent, in terms of the molecules of which they are constructed, assuming them to behave according to the laws of physics and chemistry as exhibited in inanimate matter. In this book I have attempted to give in outline an account of this development up to the stage reached at the present time, as far as possible in non-technical language. When technical terms cannot be avoided they are introduced in a way which I hope will be within the understanding of the non-scientist.

I have also drawn attention to the limitations of present day knowledge. This is all the more necessary as very misleading accounts of scientific discoveries are often reported in the press and on television. The uninformed reader gets the impression that nearly everything has been discovered and what remains is comparatively insignificant.

It is usual for scientists to attempt to draw the widest possible conclusions from their findings. This is entirely legitimate and is usually hedged about by suitable provisos. When reports of these findings find their way into the popular press, the qualifications are usually forgotten. The successful fertilisation of a human egg cell outside the mother's ovary and its continued life for perhaps a day or two is hailed as the successful or imminent development of a test-tube baby. A small modification of a virus is reported as implying that science is able to produce new and terrifying viruses for which no immunity could be found. The successful synthesis of a virus in a test-tube is proclaimed as the creation of life *de novo,* ignoring the fact that all the agents used in the synthesis such as the enzymes and synthetic mechanisms, as well as an original virus which is copied, are provided by living cells.

This is not to belittle the real advances made in the knowledge of living organisms and how they work, but the great difficulties encountered in advancing further in certain directions are ignored. There remain enormous tracts of biology, especially in the more complex organisms and in the field of mentality, which are still regions of ignorance rather than of knowledge.

Having given my readers, I hope, an adequate idea of what is known about how living organisms function, I turn to the question of whether free will is possible in such organisms, since it might be expected that that which occurs in a physicochemical organisation will be determined entirely by the forces acting on it. I conclude that because behaviour particularly in the higher animals and man is determined not only by the circumstances of the moment but also by recollections of the past, some free will is possible, according to the ability of the animal to learn from its past experience.

The question then arises for human beings, on what basis are we to choose between the different possibilities open to us. Can scientific knowledge provide a basis for ethics, i.e. criteria for distinguishing between what is permissible and what not? I argue that even if scientific knowledge were much more complete then it is at present it would still fail to provide the criteria of value which human life requires.

How then can we find a basis for judgements of value? While they provide an accumulation of the wisdom of past generations the different religions have become vulnerable to the advance of science. It becomes necessary in each generation to redefine the frontier between knowledge and belief. Here we are on highly controversial ground and I can only offer some thoughts which may help to clarify the situation for those, especially younger readers, who are puzzled by the apparent irrelevance of religion in a scientific world.

Although I necessarily give an account of the chemical basis of life, which is the background of the whole discussion, I do not assume any technical knowledge by my readers. Those technical terms which are unavoidable are explained as we go along in, I hope, understandable ways.

I have made use, in a revised and simplified form, of some passages from a previous book, *Science and Human Life,* which was published by Pergamon Press and Basic Books Incorp. in 1957 and is now out of print.

Rickmansworth 1976

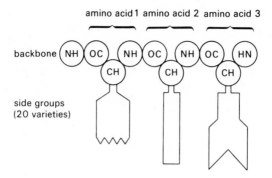

Fig. 4   A small part of a protein chain. Most protein
chains are 100 or more units long.

possible (with twenty groups available) to make 8000 (i.e. 20 × 20 × 20)
different combinations and the number increases twenty times each time
the backbone is lengthened by one unit. In a protein chain consisting of a
hundred such units the number of possibilities is enormous*. Nevertheless
it has been found that in each protein the order in which the attached
groups come along the chain is exactly the same in all the molecules. One
of the simplest proteins is insulin which has 51 units arranged in two
chains, one with twenty and the other with thirty-one units. Nevertheless
the order in which the different groups are arranged in these chains is
exactly the same in all molecules of insulin from one species†; as first
shown by Dr. Frederick Sanger at Cambridge in 1952.

All living organisms contain large numbers of different proteins. While
it was realised that these were probably the most characteristic substances
in living cells, little was known in the nineteenth century of their
functions. It was increasingly realised by biochemists that cells are the site
of intense chemical activity. They take in the nutrient materials with
which they are provided and transform them so as to produce the
substances they require to live and to grow.

It was at first thought that all this chemical activity was an ability of
the cell as a whole, but E. Buchner (1897) discovered that juices squeezed
out of yeast cells were also able to bring about the chemical changes which
occurred in life, in particular the fermentation of sugars. He therefore

*It has been calculated that the number of possible proteins which
could occur in a protein chain 100 units long is 10^{130}. This is a
colossal number, much greater than the total number of atoms in the
whole universe.

†There are slight differences between the insulins prepared from
different species.

concluded that the cell juices contain substances, which he called *enzymes,* which were able to bring about these chemical operations.

During the early years of the twentieth century, a large number of distinct enzymes, between them capable of performing an enormous variety of chemical operations, were recognised. But the nature of enzymes remained obscure for a long time although many attempts were made to concentrate them and to obtain them in a purified form. The difficulty was the comparatively small quality of each enzyme present in many tissues and to separate the large number of distinct enzymes from each other and from other cell constituents was a formidable problem for biochemists.

It was not until 1926 that really pure enzymes were obtained. Then J.B. Sumner at Ithaca, New York, isolated a protein in crystalline form from Jack beans and showed that it behaved exactly as the enzyme urease, which is able to decompose urea. Shortly afterwards John H. Northrop and Moses Kunitz at the Rockefeller Institute, Princeton, New Jersey, were able to prepare in crystalline form from extracts of the pancreas glands of animals the enzyme, tryspin, which digests proteins in the intestine. This was also a protein and though it was conjectured that the protein was only a carrier of a small molecule which had the enzyme activity, exhaustive attempts to separate the enzyme activity from the protein failed and it was concluded that the protein *was* the enzyme.

Since this time many enzymes have been purified and in every case they have been found to be proteins. This led to the conclusion that in addition to other functions, such as providing the structural parts of tissues (muscles, skin, etc.) proteins are the agents which bring about chemical processes in cells.

Proteins with enzyme activity exist in great variety in all living cells, and it is estimated that most cells probably have a thousand or more distinct enzymes. It is a remarkable fact that micro-organisms such as bacteria contain enzymes which are very similar to those which are found in the cells of higher organisms. In fact it looks as if the majority of enzymes required in living processes were produced at an early stage in the development of life, perhaps before any complex forms of life existed. Indeed bacteria often possess a greater range of chemical ability and therefore more enzymes than the cells of higher organisms, since they can exist in a very simple environment, while animals require a complex and varied diet.

How is it that proteins have these many and varied abilities of bringing about chemical reactions? This question can only be answered in general terms at present. The sequence of the amino acid groups is known only for comparatively few enzymes. The arrangement of the chains in space is known in even fewer cases, but the number is rapidly increasing. This arrangement has been demonstrated by studying the X-ray diffraction from protein crystals—the interpretation of which was first achieved by

Fig. 5   Shape of the
          myoglobin molecule.
          Myoglobin, a protein
          similar to haemo-
          globin, carries the
          oxygen in muscles.
          The actual protein
          fibre is coiled inside
          the sausage—like
          tubes. (Adapted
          from a diagram by
          R.E. Dickerson,
          California Institute
          of Technology.)

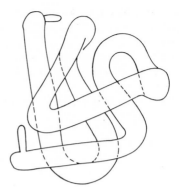

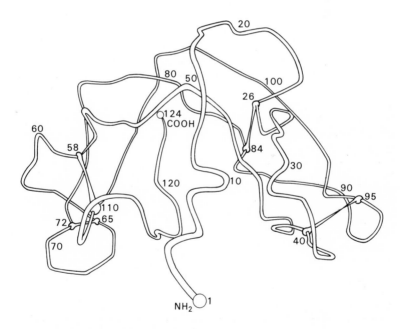

Fig. 6   Shape of the protein fibre in an enzyme, ribonuclease,
          which can digest ribonucleic acids. The numbers refer to
          the positions of the amino acids as they occur along the
          fibre. There are 124 amino acid units in the whole fibre.
          The links between positions 26-84, 58-110, 65-72 and
          40-95 are -S-S- bridges, provided by the amino acid cystine.
          (Adapted from a drawing by G. Kartha, J. Bello and D.
          Harper.)

Max Perutz and John Kendrew at Cambridge in the early 1960's with the blood proteins haemoglobin and myoglobin. In these it was found that the protein threads which are flexible are bent into complicated shapes (Figs. 5 and 6). This brings amino acid groups into close proximity which may be widely separated in the serial sequence in which the groups occur in the protein chain.

The amino acid groups themselves are quite varied in character. Some are acidic, others basic etc., and the result of the bending is to bring together a number of groups which provide regions in the protein (called active centres) which have a special character and which provide an environment which is favourable for chemical changes of various types.

The ability of a protein to act as an enzyme is thus determined very precisely by the shape its protein chain takes up and this is exactly determined by the order of the amino acid groups in the protein chain. In order to make particular enzymes, life must therefore have found a way of constructing protein chains with amino acids in a precise order — and this order is precisely repeated every time an enzyme molecule is made.

This is now recognised as the characteristic ability of life in all its manifestations—the ability to construct large numbers of different proteins, each having a precisely determined order of amino acid groups. Since the characteristics of an organism are transmitted from one generation to the next by means of the genes, it follows that the primary task of genes is to specify the order of amino acid groups in protein chains so as to produce distinct enzymes. How this is achieved I shall explain in the next chapter.

4 The heart of the life process

We come now to the second important class of giant molecules found in living cells—the nucleic acids. First discovered by Friedrich Mieschler at Tübingen in 1869, they were soon recognised as important constituents of chromosomes (which contain both nucleic acids and proteins). Later it was realised that the nucleic acid of chromosomes was the material of which genes are constructed—a fact confirmed by the discovery that in certain cases it was possible to change bacteria from one type to another type by merely adding nucleic acids extracted from the second type to cultures of the first type.

It was found that chromosomes had a special kind of nucleic acid which was named DNA (short for deoxyribonucleic acid, after the special sugar, deoxyribose, it contained). Another kind of nucleic acid, called RNA (ribonucleic acid) was found to be present especially in the cytoplasm of cells (i.e. the parts outside the nucleus).

These nucleic acids were found to be present in all forms of life—plants and animals, as well as in micro-organisms. Intensive studies were made of their composition and it was found that DNA was made up of smaller units called nucleotides. Each of these units consists of a basic substance combined with the sugar (deoxyribose) and a phosphate group. In DNA the sugar and phosphate groups are joined alternately so as to form a continuing backbone (Fig. 7) from which the bases project at right angles

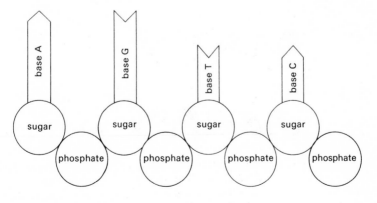

Fig. 7   Chemical structure of a DNA fibre.

to the direction of the fibre. The bases are of the four kinds, labelled A, T, G, C respectively. The outline chemical structures of these are given in (Fig. 8). These bases are very flat molecules which can be packed tightly at the side of the sugar-phosphate backbone forming a tight column like a pile of coins.

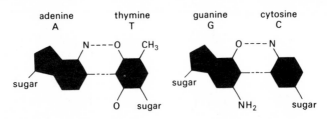

Fig. 8   The bases A and T, G and C, showing how they
           combine together in pairs.

It was discovered by Erwin Chargaff of Columbia University, who made analyses of numerous samples of DNA, about 1950, that the amount of A was always nearly equal to that of T and the amount of C to that of G. The reason for this came to light when James Watson and Francis Crick at Cambridge in 1953 noticed in the course of building molecular models of DNA, that A and T and also G and C had complementary structures which would fit together in a unique way. As a result of this they found that a regular model of DNA could only be made if the units were arranged in twin threads wound round each other, in such a way that whenever group A was present on one thread the group T occurred on the other—similarly C in one thread was bound to G in the other (Fig. 9). This structure was confirmed by the X-ray diffraction pictures of Maurice Wilkins at King's College, London.

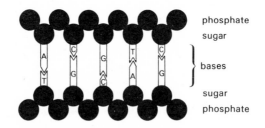

Fig. 9   Double stranded structure of
           DNA. In DNA, as obtained from
           cells, two strands are wound
           round each other in the form of a
           spiral as in Fig. 10.

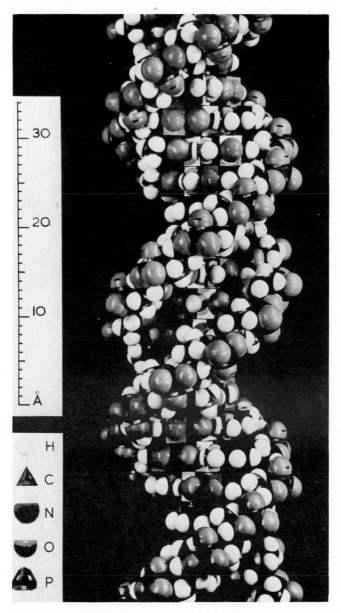

Fig. 10   Crick and Watson's model of a DNA fibre. The spirals
are the sugar-phosphate backbones. Between these are
the bases which provide the genetic code.)Repro-
duced by courtesy of Professor M.H.F. Wilkins.)

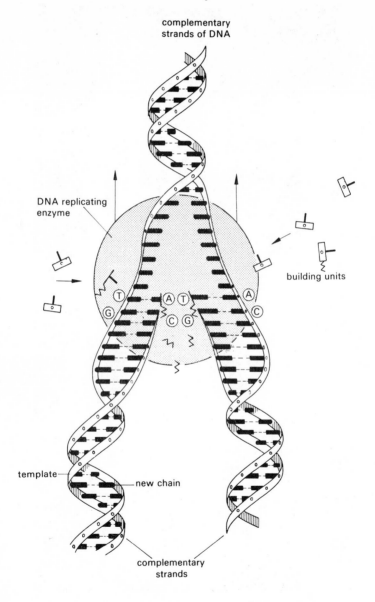

Fig. 11   How genes are replicated. Each strand of DNA acts as a
          template for the formation of a complemetary strand.
          Adapted with permission from E.J. Ambrose and D.M.
          Easty, *Cell Biology,* Nelson, 1970.)

It was immediately recognised by Crick and Watson that if DNA was in fact the genetic material, then this provided an explanation of how replication may occur. They realised that if the two threads of the complete DNA fibre are separated, a complete DNA fibre can be built up on each of them in one way only if it is to retain the matching of A to T and C to G. If this is done we then have two DNA (double stranded) fibres instead of the original one (Fig. 11). This observation provided genetics with its basic theory, viz. an explanation of how, if genes are composed of DNA, they can undergo replication. This was a remarkable advance in genetic theory but it did not immediately demonstrate how the genetic message carried by the DNA could control the nature of the organism. If the function of the genes is to give information which primarily determines the order of the amino acid groups in the proteins, it was realised that the order in which the bases A, T, C, G come on one of the DNA fibres must provide a code which specified the order of the amino acid groups in the protein. But this presented a difficulty; there are twenty different amino acid units, but only four different units in DNA.

The problem is how a four letter alphabet can be used to provide a code for the positions of twenty different units in a protein chain. The answer was not immediately obvious, but it soon became apparent that more than one of the code letters would have to be used to specify the position of each amino acid unit. From four letters sixteen combinations of two letters can be made, which is not enough to specify 20 amino acids. Taking three letters at a time we have a possible $4 \times 4 \times 4 = 64$ combinations at our disposal, which provides more than enough combinations. Thus the sequences GCA, CAG, AAA as carried by one of the strands of the DNA are some of them. These are the codes for the amino acids arginine, valine, and phenylalanine respectively.

It has been amply demonstrated that the code employed is in fact a triplet code and the triplet groups of code letters which stand for all the amino acids are now known, mainly through the experiments of Dr. Marshall Nirenberg at the National Heart Institute, Bethesda and Dr. H.G. Khorana, at the Institute of Enzyme Research, Madison, Wisconsin. They were able to identify the codes for all the 20 amino acids which occur in proteins by using synthetic nucleic acids with known sequences of the letters. The code is the same, so far as it has been examined, for all organisms—one more proof of the essential unity of all forms of life.

This still leaves the question of how the code carried by the DNA is actually used to guide the construction of proteins. It was found that proteins are actually made by small particles called ribosomes which occur in large numbers in most cells. The code as carried by the DNA is not used directly to guide the synthesis of protein. It was found that it was first transferred to an intermediate substance—a form of messenger RNA, first recognised by Dr. F. Gros of the Pasteur Institute Paris, which carries the

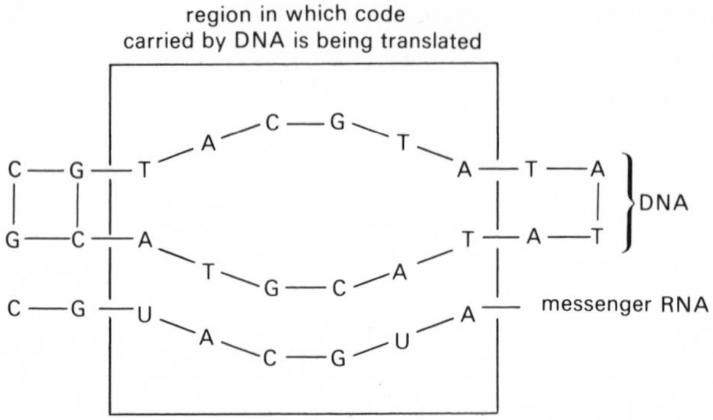

Fig. 12   Translation of the code of DNA into messenger
RNA.

code from the DNA to the ribosomes, where it is read and 'translated' into protein (Fig. 12).

Messenger RNA is similar to one strand of the DNA except that it has a different sugar, ribose instead of deoxyribose, and instead of thymine (T) it contains a very similar substance, uracil (U). It is formed by a process which is very similar to the replication of DNA. A portion of the DNA spiral is unwound and on one of the strands a complementary strand of RNA is built i.e. where there is G on the DNA, C will be inserted in the RNA and where there is A on the DNA, U will be inserted in the RNA. The code as carried by the DNA strand which is being 'translated' will thus appear in the messenger RNA in a complementary form, as in Fig. 12.

The process of synthesising the protein can now begin. A ribosome attaches itself to the end of a fibre of messenger RNA and as it moves along it a protein is formed according to the message carried by the code. Many details of how this is achieved are still obscure, but probably the most important features are known, as a result of the efforts of numerous investigators (Fig. 13).

Ribosomes occur in great profusion in all living organisms—both simple and complex They are to be found in every type of life—from the simplest bacteria to the most complicated animals. They are at the centre of the life phenomenon. They operate at a very high speed. It has been estimated that a protein chain containing 1200 amino acids is made by the *E. coli* bacterium in 25 seconds i.e. about 50 amino acid units are added to the growing protein chain by a single ribosome in each second. We thus find mechanisms of a most complicated and efficient kind at the heart of the life process.

Although still there are features of bacterial life which are not understood this provides a very impressive demonstration that the simplest

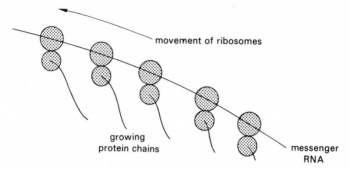

movement of ribosomes

growing
protein chains

messenger
RNA

Fig. 13   Synthesis of protein by the ribosomes as instructed
by the code on the messenger RNA. The ribosome
consists of two parts, one of which 'reads' the
message of the code and the other inserts the amino
acid units into the protein chain according to the
message. The protein chain thus increases in length as
the ribosome moves along the messenger fibre. There
may be several ribosomes operating at the same time
on one fibre of messenger RNA.

forms of life, such as bacteria, are in fact chemical machines, whose
behaviour can be explained by the chemical molecules which they contain.
However, the picture of even unicellular organisms, such as bacteria, is
already one of considerable complication. The substances of which
bacteria are composed are contained in a membrane or bag which is itself a
fairly complex structure. It will admit the food materials which the
organism requires and it will pass waste products which are not required in
the reverse direction.

Inside the bag are all the enzymes required to bring about the
complicated processes which the cell has to perform in order to live. These
involve breaking down the food materials into suitable usable products and
building them up into all the complex compounds which are required for
growth—especially proteins and nucleic acids. It is probable that simple
bacteria use at one time or another more than 1000 different enzymes,
although the total number has not yet been established.

Besides the enzymes, the cell contains the enzyme factories, the
ribosomes, which have already been described and also various accessories
such as 'transfer RNAs', at least one of every kind of amino acid which
help them to find their places on the growing protein chain according to
the code. It also contains the DNA of the cell, which carries the codes
which enable all the other cell constituents to be constructed. In addition
to the messenger RNAs, which, as mentioned above, transmit the message
carried by the DNA to the ribosomes, where it is used for the manufacture
of proteins, the DNA also provides the codes for the construction of the

proteins of the ribosomes and the many other proteins which the cell requires such as the enzymes which are required in the synthetic process itself and also those which are involved in the replication of DNA itself. The bacterial cell is thus a remarkably self-contained system of activities, in which every part is necessary for the functioning of all the other parts. The DNA thus provides the codes for making the apparatus required for its own replication and for its functioning. It can be seen that the DNA is paramount; everything else in the cell is derived from it.

How much DNA is required to carry the codes for all these enzymes and other proteins required for the operation of the simplest kind of cells? The common bacterium present in the colon, *Escherichia coli*, which has been investigated more extensively than any other micro-organism, has in its DNA about six million pairs of base units, which is sufficient to provide two million coding 'triplets'. This will make a thread which when extended would be about 3 millimetres long. As the bacterium is much smaller than this, being only 1/500 of a millimetre across, the DNA thread must be folded many times inside the bacterium. If we suppose that an 'average' protein contains at least 100 amino acid units, each such protein will require in the DNA to code for it, 100 'triplets' i.e. 300 code letters. There is thus in each cell of *E. coli* enough DNA to code for about 20 000 different 'average' proteins.* There is evidently enough DNA to provide genes for all the proteins required, with probably a good many to spare. It is of course possible that more than one example of some genes is present in the bacterial chromosome. The bacterial cell, one of the simpler manifestations of life, is clearly a marvel of complex organisation. All the mechanisms required for life are packed into what is, from our point of view, a very small volume, forming an organism which under suitable circumstances, is capable of multiplying itself indefinitely (Fig. 14).

Mutations

If the DNA in the cell is paramount it follows that any agency which modifies the DNA of the cell will bring about changes of the constitution and performance of the cell itself. A change in the nature of an organism which can be inherited by the descendents is called a mutation. There are in fact numerous ways in which the DNA complement of cells can be modified. For many years it was held that the genes are unalterable by outside agencies and only in 1927 was it discovered by Dr. H.G. Müller. at Madison Wisconsin, that mutations could be produced by exposing organisms such as fruit flies to X-rays and similar radiations and later, in 1936, it was found by Drs. Charlotte Auerbach and Robson in Edinburgh that exposing the germ cells of organisms to certain chemical substances also

*i.e. 6 000 000/300= 20 000

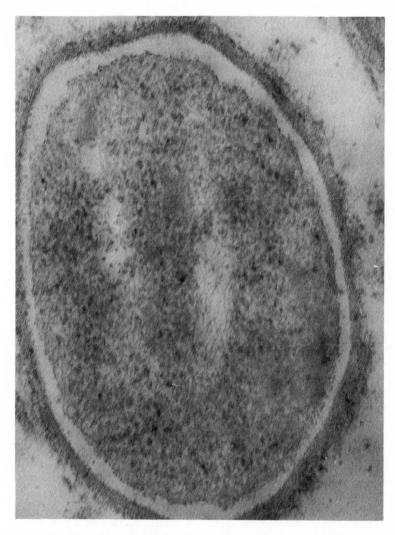

Fig. 14   The inner world of a bacterium. Cross section of *B. megaterium.*
The few fibres seen near the centre in the lighter areas are DNA
fibres, which constitute the bacterial chromosome. (By Drs. E.
H. Mercer and M.S.C. Birkbeck.)

causes mutations to occur. An enormous amount of investigation has been
done on such effects; particularly with micro-organisms, which can easily be
subjected to such agents. In these cases millions of individuals can be
exposed at the same time to the mutagenic agent, (i.e. an agent which
produces mutations) and it is possible technically to detect rare effects,

even if they only occur in one in every million or more of the organisms which are treated.

These studies have contributed very greatly to the knowledge of how heredity works in micro-organisms. It was found that in most of the mutations produced by these agents, there was a loss of the ability to perform just one chemical reaction. It was inferred from this that the mutated organism had lost the ability to make one enzyme and therefore that one gene had been damaged so that it was unable to provide the correct code for this enzyme†.

From what we know of the nature of the genetic code we should expect that any change in the nucleotide units of the DNA may be sufficient to modify the code and thus to upset the protein which is coded for. The destruction of a single unit may cause a coding triplet to be read wrongly so that a wrong amino acid is inserted into an enzyme molecule. Sometimes this may be unimportant, as when the amino acid is replaced by a very similar one, or if it happens to be at an uncritical point of the protein structure. But if the amino acid is changed at a critical point of the enzyme, the latter will not function properly and if it is an important enzyme the organism will then be unable to live. In other cases the organism may still be able to live and multiply but its characteristics, such as its food requirements, may be different.

Many hundreds of mutated organisms have been produced artificially by exposing bacteria and other organisms to mutagenic agents such as X-rays, radiations which are produced by radio-active substances and certain chemical substances. These are usually deleterious to the life of the organism and result in a limitation of the conditions under which the organism can live.

When the genes of an organism are exposed to a mutagenic agent, it is a matter of chance which gene or portion of a gene is affected. Usually the nature of the mutation which is brought about cannot be controlled*. Usually as mentioned above mutations produced by mutagenic agents are the result of a loss or damage of genes. Processes have also been discovered by which bacteria and their cells can gain new genes.

Thus, it was discovered in the course of investigations on micro-organisms that mechanisms exist by which genes can be exchanged between different cells. If we have two batches of a micro-organism, one of which has lost one gene and the other has lost a second gene and we grow them together, it is found that a few organisms are produced which are complete in the sense that they contain both of the missing genes. This

---

†With larger doses of the mutagenic agent, two or more genes may be lost at the same time.

*Some chemical mutagenic agents may have a greater ability to produce some mutations than others.

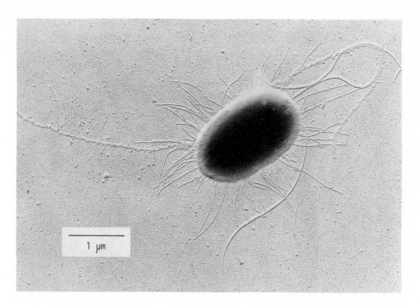

Fig. 15   A cell of *E. coli* showing sex pili and flagella. The sex pili are the short straight fibres through which transfer of genetic material to 'female' cells can take place. The longer flexible fibres are flagella which by their undulations can propel the cell through the medium. (Photograph by Professor M. Tomoeda of Kanazawa University, Japan, with whose permission it is reproduced.)

means that a cell which has a complete complement of genes has been formed from two defective ones. This phenomenon (recombination) occurs quite frequently among simple organisms. Indeed some bacteria have a special mechanism which enables them to exchange genetic material. This was discovered by Drs. E.L. Wollman, F. Jacob and W. Hayes in 1953 at the Pasteur Institute in Paris, who found that certain strains of the bacterium *E. coli* had cells of two types, one type smooth (called female cells) and the other rougher (male cells). The latter had little tubes on their surfaces, which made contact with the smooth (female) cells and enabled DNA to be transferred from the male to the female cells. It seems that sex occurs even in bacteria. In fact an extensive exchange of genetic material may occur between different cells. Not only can defective sets of genes be repaired, but mutants which have acquired new and advantageous genes may be able to transfer them to other cells. It must be remembered that the process does not stop with the formation of a single 'recombined' cell—if the new cell is viable it may be replicated many times with its new complement of genes and in fact may constitute a new strain of the organism which replaces the original strain (Fig. 16).

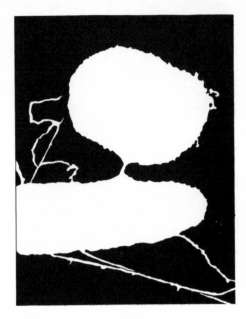

Fig. 16   Two cells of *E. coli* in conjugation. (After Wollman, Jacob and Hayes.)

As already mentioned, DNA by itself is also capable of entering cells and becoming part of their genetic complement. This was discovered by Drs. A.J. Avery, C.M. McCleod and M. McCarty of the Rockefeller Institute, New York in 1944 who found that when the DNA from one type of cell was prepared and added to growing cells of a second kind, characteristics present in the first cells were acquired by the second type and were also transmitted to their descendants. In such cases a cell transformation involving a change of genes had been brought about by the added DNA. For example, as mentioned above there were two types of pneumonia bacteria—called rough and smooth, both of which when replicated, remain true to type. But adding DNA from the rough type caused the smooth cells to be transformed to rough.

A similar phenomenon occurred with certain bacteria in which some cells were sensitive to the antibiotic penicillin while others were resistant to it, because they possess a gene which makes an enzyme which destroys penicillin. Addition of DNA made from penicillin resistant cells to the sensitive ones again enables the latter to produce the penicillin destroying enzyme and so become resistant.

Yet another way in which the genetic character of cells can be modified is when particles of virus enter the cell and under some circumstances part of the viral DNA may become attached to the chromosome and so provide additional genes. To explain this we must first give some account of the viruses.

5 The smallest manifestations of life

Before we consider the more complex forms of life it will be useful to look at the very simplest 'organisms' which exist. Although they are not far removed from simple chemical molecules—in fact in a few cases their chemical structures are almost completely known—they possess two of the fundamental abilities of all life; viz., they are able to replicate themselves in a suitable environment and to provide the necessary codes for constructing their own proteins.

They are the viruses, agents which are much smaller than bacteria, which cause diseases in plants and animals. They are invisible in optical microscopes, but they can be made visible in the more powerful electron microscopes. Many viruses have been isolated and investigated during the last forty years and a great deal of insight on the basic processes of life has been obtained from them.

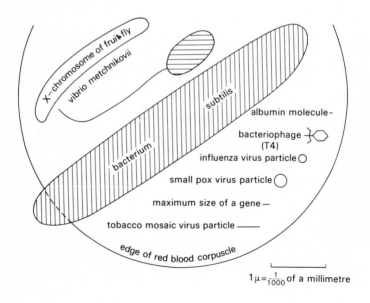

Fig.17   Relative sizes of virus particles and bacteria. (Compare with Fig.00).

They come in a great variety of shapes and sizes (see Figs. 17 and 18). Some virus particles were found to be spherical or rhombohedral, others are elongated cylinders or like tadpoles, a compact body with a tail or tails. They all contain a nucleic acid: some contain RNA, as is the case with tobacco mosaic virus (which causes a mottling of the leaves of tobacco plants) and the viruses of mumps and measles; others contain DNA (such as herpes and smallpox). In addition to a nucleic acid they all when isolated contain some proteins, which have various functions, such as enabling the virus to penetrate cell walls or to provide a protective capsule for the virus, when it is out of the cell.

The nucleic acids, which often form a closed loop when released from the virus particles, are the essential part of the virus, because by themselves they can cause infection; in fact, the nucleic acids carry the codes required

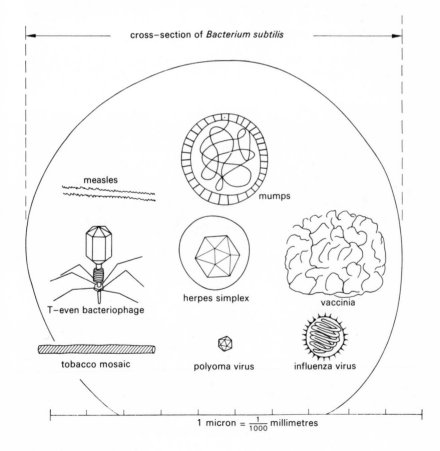

cross-section of *Bacterium subtilis*

measles

mumps

T-even bacteriophage

herpes simplex

vaccinia

tobacco mosaic

polyoma virus

influenza virus

1 micron = $\frac{1}{1000}$ millimetres

Fig. 18   Some viruses.

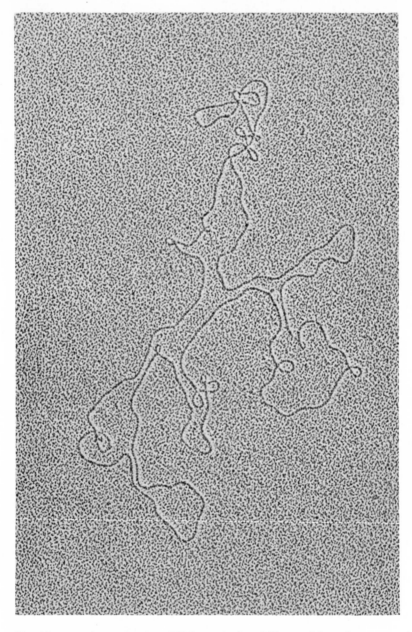

Fig. 19   DNA fibre of a virus (T3 bacteriophage) liberated from its
capsule. (by D.A. Ritchie, C.A. Thomas, Jr., L.A. Machattie and
P. C. Wensack.)

to make the virus proteins. In the case of DNA viruses the code is transcribed into a messenger RNA before being translated by the usual protein synthesising mechanism (ribosomes etc.) of the cell. In the case of RNA viruses, the RNA itself acts as its own messenger and guides the formation of the proteins required.

Viruses can normally only function inside living cells, where they make use of the synthetic machinery e.g. ribosomes, and other parts of the synthetic mechanisms, which they find there. They are therefore not complete living organisms, since they are dependent on living cells. Nevertheless they are of considerable interest as intermediates between the living and the non-living, which possess some of the abilities of life.

Some viruses are quite small, especially the group of viruses which attack bacteria, known as bacteriophages. One example of this type of virus, is one known as R17. The whole particle contains only about 3000 of the usual nucleotide 'units', in this case RNA, which are sufficient to provide codes for only 1000 amino acid units, which could form from three to ten protein chains depending on their size. Three of these proteins have been identified: one is the coat protein which provides a capsule to protect virus particles when outside the cells; the second is an enzyme which brings about the replication of the viral RNA (the bacterial cell itself does not have an enzyme which can do this). The third protein provides a cap which completes the capsule.

The RNA of this and similar viruses is not unmanageably large for chemical analysis and it may soon be possible to determine the complete sequences of their nucleotide units. A start has been made for example on a similar virus, known as Q_β, by a group of biochemists in Bristol (Hindley, Billiter and colleagues) who have determined the order of the first 175 nucleotide units of the RNA of this virus. The sequence of nucleotides on a portion of the RNA, 387 units long, of the virus known as MS2 which forms the gene of the coat protein, has also been determined (1972) by three Belgian investigators. It can only be a matter of time before the whole sequence of these viruses is determined. This means that we have the prospect of completely specifying a virus as a chemical molecule. The total chemical synthesis of the virus would involve assembling 3000 or more of the nucleotide units in a unique order—a task of great difficulty by the methods of synthetic chemistry (unless there are repeating patterns, which is not impossible) but one which is carried out in life at a great rate.

But may it not be possible to persuade the virus to replicate itself outside a living cell? This ought to be possible if the virus is provided with the mechanisms required for its replication, such as the synthetic enzymes and the necessary small molecules required for its construction. A successful attempt to achieve this was made in 1967 by Drs. Goulian and Kornberg (both of Stanford University) and Dr. Sinsheimer of the California Institute of Technology, with the virus QX174—a virus containing DNA as its nucleic acid. These investigators found that when the viral DNA was incubated with the DNA replicating enzyme and the necessary small molecules (nucleotides), complete new viral DNA molecules with demonstrable viral activity could not be detected although some DNA synthesis took place and small sections of DNA were formed. It was found that another enzyme was required, a 'bridging' or 'joining' enzyme, which joined up smaller pieces of DNA made by the DNA replicating enzymes and when this was added, the complete viral DNA was formed, which was found to be biologically active.

This was hailed in the popular press as the artificial synthesis of a living entity. But in so far as use was made of natural enzymes derived from living cells and also some previously existing virus as a 'template', it obviously falls far short of this—although it was a remarkable achievement, in that the living process had been broken down into its elementary parts, which by themselves are capable of being carried out in test tubes.

Another remarkable property of some DNA viruses is their ability to enter their host cell and remain there apparently dormant for long periods, in fact through a number of generations during which cell division has taken place. It was shown by Drs. Jacob, Lwoff and collaborators in Paris in 1953 that in these cases (called lysogenic viruses) the viral DNA becomes part of the bacterial chromosome and when the chromosome replicates the viral DNA replicates also. These viruses may be quite large; thus the one investigated by Jacob and Lwoff has about 50 000 nucleotide units i.e. a sizeable piece of DNA is added to the chromosome when the virus is incorporated. It may remain as part of the chromosome for a number of generations, but under some circumstances (e.g. some shock such as exposure to ultraviolet light) the virus portion may be liberated from the chromosome and it then behaves as an active virus again, replicating in the cell to give new virus particles and so becoming 'virulent'.

This process provides a useful method of adding genes carried by virus particles into the gene complement of cells. The reverse is also true, because when the virus particle is released from the cell it may take with it one or more genes belonging to the cell. In this way viruses may acquire extra genes which may sometimes be useful to them. It is probable that new virus strains may originate in this way by the capture of genes from their host cells.

These experiments give us a picture of the bacterial DNA not as an unalterable collection of genes, but as a set of genes which may receive additions from without and lose others. It is also possible that a part of the gene complement may separate from the chromosome and attain a life of its own as a virus. Possibly viruses originated in this way, but not every separated piece of DNA or of RNA could act as a virus. An effective virus must consist of genes which code for very special proteins e.g. which are capable of providing a capsule to protect the viral nucleic acid when the virus is free and also to provide a means of entering and leaving the host cell. For example some viruses carry the enzyme lysozyme which enables them to dissolve the cell wall of the host cells and so to enter the cells.

While it is theoretically possible for any piece of DNA to act as a virus i.e. to make use of the synthetic machinery of the cell and to produce the proteins for which it codes, only certain highly specific proteins will have any useful function.

The suggestion that 'biological engineers' of the future may be able to synthesise new and terrible viruses for which no remedy will be possible, is an exaggeration. The known viruses must have arisen over a long period of evolution and are superb examples of precise adaptation to the circumstances in which they occur. This is not to say that new viruses cannot arise. In the course of its sojourn in its host cells a virus may pick up or lose various genes and so change its genetic constitution. Also the genetic constitution of the host cells may also change by incorporating various viruses or parts of them, and as a result we must see the bacteria and viruses not as unchangeable species but in a state of constant change. In fact it has been suggested that the genetic complement of many organisms has been formed by the coming together of a variety of genes which have had different origins.

The rules of genetics as applied to bacteria apply equally to viruses. Their genes may undergo mutations; most of which will probably be harmful to the virus, but occasionally a mutation may occur which is advantageous. New strains of a virus frequently occur as the result of such changes and may spread because no effective immunity to them has been developed in their hosts.

While this may occur spontaneously from time to time, we must remember that viruses must have existed for a very long time—perhaps for as long as cells themselves—and every possible mutation may have already occurred at some time or another and those which have survived are those which are best fitted to survive in the forms of life which now exist. Nevertheless it is recognised that the situation has its dangers. For example, it is possible to isolate from animal and other cells genes which confer resistance to penicillin or other antibiotics and if these are introduced into certain bacteria a resistant strain of the bacterium will be produced. Also there are viruses which cause the development of tumours in some animals (this has never been proven to occur in human beings). Genes from these viruses can be introduced into bacteria and there is a danger that infection with such bacteria would result in tumours in humans. Such consequences are uncertain, but in view of the possibilities a group of U.S. molecular biologists have urged that experiments involving the introduction of animal genes into bacteria should be suspended until a better assessment of possible hazards can be made and the U.S. Academy of Sciences has endorsed this proposal (1974).

The British authorities have also adopted a somewhat less stringent code of experiments. One suggestion is that such experiments should only be done on strains of bacteria which are unable to survive in animals.

The ease with which new genes can be inserted into bacteria and even into animal cells has suggested that similar processes might be used to cure some human diseases which are due to the absence in the patient's chromosomes of the gene for an enzyme which carries out an important step in a series of reactions. As a result the individual is unable to complete this reaction sequence and a chemical substance, which is produced in the reaction sequence, accumulates because it is not converted in the next step in the sequence, with unfortunate results. For example, a condition known as phenylketonuria which occurs in about one in 18 000 live human births, is due to lack of the enzyme which converts phenylalanine into tyrosine and instead it gets converted into toxic substances, which cause mental retardation. A similar disease, believed to be due to lack of the enzyme, arginase, which keeps down the amount of arginine in the blood, has been treated with Shope's papilloma virus—a virus which causes skin tumours in rabbits, but does not appear to harm humans—because it was believed to contain the arginase gene. However this kind of therapy is obviously fraught with difficulties and dangerous possibilities—although risk of viral infection can be reduced by 'inactivating' the virus with ultraviolet light before injection. It is difficult to ensure that the virus reaches the cells concerned, or, when there, enters them and becomes incorporated in their chromosomes. Its presence may also produce unwished-for complications. A more hopeful possibility may be to isolate and inject the missing gene itself e.g. as obtained from bacteria. It may become possible to prepare genes in quantity from living cells or replicate them by the use of synthetic enzymes and inject them directly in the hope that they will enter the defective cells and become incorporated in their chromosomes. This is perhaps a less dangerous procedure but the possible complications are completely unknown. 'Gene therapy' and 'genetic engineering' are thus at a very primitive stage of development and it is doubtful if they can be used effectively in the present state of knowledge or for many years in the future.

6 The origin of life

Until well into the nineteenth century, it appeared obvious that living organisms differed in essential respects from inanimate matter, and this led to the view that they were infused by a 'vital principle', which enabled them to be alive. The steady progress of science in other directions, long before any substantial development of biochemistry, led scientists and philosophers to speculate that in time organisms would be brought within the same natural laws as inanimate objects, but this was more an expression of faith in the uniformity of nature than a conclusion from a sound basis of knowledge.

The great discoveries of the twentieth century of the detailed mechanisms of life, which have been outlined in the earlier chapters, have amply justified this faith. It is no longer possible to doubt that living organisms are, in fact, composed of chemical substances of definable chemical composition and although scientific explanations of their behaviour are still incomplete, most of the known facts can be interpreted in terms of the laws of chemistry and physics. Accordingly there is no need to invoke any special 'vital principle' to account for their behaviour. If the atoms and molecules could be assembled into the same 'macro-molecules'—an enormously difficult task on technical grounds, but not impossible in principle—there is no doubt that a living organism would result, which could go on replicating itself in a suitable medium. There thus appears to be no difficulty in principle in creating life in a test-tube—although the practical difficulties of such a programme are immense.

How then did life succeed in organising itself on an earth which was originally lifeless? The fact that all forms of life use similar basic mechanisms suggests that living things had a single origin from which the innumerable forms of life present today are descended.

The ascent of life can in general only be traced a little way back by means of the fossil record. Until hard and permanent structures were developed there was little to leave a permanent record. Most of the essential features of life must have been present long before the fossil record begins. Nevertheless 'fossil' remains of what appear to be algae have been found by Drs. E.S. Barghoorn and S.W.E. Schopf of Harvard University in certain carboniferous slates which are between two and three billion years old. This would suggest that a fairly complex cell life had been elaborated more than about 3 billion years ago i.e. quite soon after the formation of the earth, according to present measurements.

At its minimum life involves two different cycles of activity, one involving the nucleic acids and the other involving proteins. But as we have seen these are not independent of each other. The replication and synthesis of the nucleic acids is brought about by protein enzymes, while the synthesis of proteins requires the codes carried by the nucleic acids and also the ribosomal synthetic machinery of which RNA is an important part. No part of this mechanism can operate without the other parts and it is difficult for us to see how much a complex arrangement could have come into existence spontaneously from simpler parts, because none of the elements of this machine could have functioned by itself and no simpler machinery which will perform the same function is known to exist or can be easily visualised.

Although we have very little idea how such complex molecular systems could have originated spontaneously, there is ample evidence that organic compounds which are the building blocks from which these structures could be elaborated, were present in large quantities in the primitive earth. It was suggested by G.I. Oparin, a Russian biochemist, that life could only have originated in an environment which was already rich in organic compounds. This original environment has completely disappeared and most organic compounds at present on the earth are themselves the products of life*. Oparin thought that ultraviolet light and electric discharges in the early atmosphere, which is believed to have been deficient in oxygen, would have given rise to simple compounds such as ammonia and urea and simple amino acids. The fact that the latter are formed when electric sparks are passed through mixtures of carbon dioxide and ammonia has been confirmed experimentally by Dr. S.K. Miller and Professor Urey of Columbia University. But the concentration of such compounds which can be obtained is small because electric sparks also decompose the compounds formed.

However more recently numerous chemical processes have been found which could have occurred in the crust of the primitive earth and which may have produced enormous quantities of organic compounds. For example, many metals react with carbon at high temperatures to form carbides and when the earth cooled down sufficiently for the water vapour to condense, these would react with water to form acetylene and other hydrocarbons. Similarly in the presence of nitrogen, hot metals formed nitrides which can react with water to yield ammonia, and with carbon form cyanide. Cyanides and ammonia when heated together form amino acids, as Dr. Roy Markham and colleagues at Cambridge have shown and these may further condense to form primitive protein-like substances.

---

*It is possible that the enormous deposits of hydrocarbons which occur in many parts of the world are relics of the original environment, but most geologists think they are derived from decaying vegetation.

Acetylene and cyanates heated together form cytosine—one of the constituents of nucleic acids. Other nucleic acid components are formed in similar reactions. Metallic hydrides can also be formed by the action of hydrogen on hot metals and in the presence of water and carbon dioxide, formaldehyde is formed which can condense to yield sugar-like substances.

Mineral phosphates are a common constituent of the earth's crust and the formation of organic phosphates, such as occur in the nucleic acids presents no great difficulties. By the action of heat mineral phosphates are converted into 'polyphosphates' i.e. the mineral atom combined with a row of several phosphate residues. From these, organic polyphosphates could be formed, which are well known to be 'energy rich' and capable of facilitating the synthesis of phosphate polymers like the nucleic acids.

There is thus no difficulty in believing that great quantities of organic substances, which are capable of acting as precursors of the characteristic molecules of life, were present on the lifeless earth. Condensations of such molecules to give protein and also nucleic acids can easily be envisaged and experimental evidence for the formation of protein-like materials, when amino acids are heated together has been obtained by Dr. Ponnamperuma and colleagues in the Exobiology Laboratory of the Ames Research Centre of NASA in the U.S.A.

To go from this to the highly specific proteins and nucleic acids, required by the life process, presents very great difficulties and we have no idea at present how protein or nucleic acids with the units arranged in a particular and repeatable order could have been formed.

The highly sophisticated mechanisms employed today by all living organisms assume that the highly specific arrangements which constitute the genetic codes are already available, and we have no idea as to how the genetic code was established. There have been attempts to show that certain amino acids are naturally akin in shape to certain groupings present in DNA but these do not carry conviction and have not been generally accepted by scientists. It is possible that some such natural association occurred between amino acids and certain groups of nucleotides which provided a primitive code, and led to the repeated synthesis of a particular protein-like group of amino acids.

Out of a vast number of protein molecules originally formed, probably a few would have enzymic properties and some would even have the ability to assist the synthesis of nucleic acids. It is not clear how the continued formation of any useful combination could be favoured. We really have no idea how a group of proteins with such properties became associated with some nucleic acids to form a first primitive cell. This was the crucial step in the ascent of life. Was it a rare and unlikely chance which brought together the necessary elements of life, a fortunate accident which only happened perhaps once in many millions of years, or is there some principle of which we are ignorant which led to the assembly of the necessary elements in the primitive cell?

At the present time the whole of animal life is dependent on the photosynthetic activity of green plants and for this reason it has sometimes been thought that the first organisms must have been photosynthetic i.e. they were able to use the energy of sunlight in order to make the compounds which are used for building plant structures, on which animal life depends. However the photosynthetic apparatus which occurs in green leaves is not at all simple. It involves highly complex particles, called chloroplasts, which carry the green pigments to absorb light. These perform the initial steps of the synthetic process The subsequent steps of the processes used in elaborating carbohydrates involve the use of numerous enzymes so that it is difficult to see how these processes could occur until the apparatus for protein synthesis, which as we have seen involves the whole life process, was already established. No simple process of photosynthesis, not involving such an apparatus has been discovered, so that it is probable that the utilisation of sunlight was not an early stage of life but happened after the main stages of the life process, which permitted the elaboration of complicated proteins, were already in operation. Perhaps photosynthesis only occurred on a large scale when the original stock of energy-rich (prebiotic) compounds was approaching exhaustion.

It is evident that it is much more difficult than was once thought to account for the origin of life on the primitive earth. If it were only 'a fortunate accident', it is possible that life may be rare even when suitable conditions, such as a rich organic milieu, are present. For this reason it is possible that life is much scarcer in the Universe than has been thought and that although there may be many planets possessing suitable chemical and physical conditions for life, in few of them has the necessary concatenation of circumstances which led to life on Earth been present. We are unlikely to have our curiosity as to the existence of life in the planetary systems of other stars satisfied for many years, if ever. One should not however underestimate the enormous periods of time available for the establishment of the life process—probably a billion years between the formation of the earth and the first appearance of anything resembling a primitive organism. Of such a period we can form no conception, nor can we usefully hazard a guess as to the chances of the occurrence of the chemical combinations from which the life process may have developed.

The simplest living cells require a large number of distinct enzymes for their functioning and it is probable that a large number of enzymes, in fact enzymes very similar in properties to those which occur in bacteria today must have been available or were capable of being made when the earliest living cells were formed. It is a remarkable fact that enzymes are very much the same throughout the whole gamut of life from the simplest bacteria to the most complex animals. There has been in the course of evolution no steady improvement in enzymes—only in the structures they are capable of making.

Simple self-perpetuating reaction processes quite unlike those used in present life forms may have occurred and developed during many aeons into the present forms of life—but we are unable to suggest what these were and no clues have been discovered so far.

7 Complex organisms develop from single cells

So far we have concentrated on unicellular organisms, in which all the life functions are present in each cell, which is capable of living by itself and multiplying in its appropriate environment. These organisms may seem to be a primitive form of life. Their dimensions are microscopic, yet as we have seen each cell contains several thousand genes which produce the enzymes and other proteins which carry out all the cell functions, including the replication of the whole system—a remarkable feat of organisation, which represents an almost unbelievable triumph of the life process. Yet this is only the beginning of the life achievement. The forms of life which are known to us as animals and plants are built up of enormous numbers of cells of many different kinds, all combined into a single harmoniously functioning whole.

The individual cells are themselves usually much larger than bacteria, the sizes of which are usually between a thousand million and a million million cells per cubic centimetre. On the average animal cells are about a thousand times as large i.e. between a million and a thousand million cells per cubic centimetre. A largish insect weighing one gram will thus be made up of between a million and a thousand million cells. A man weighing 100 kilograms will contain something of the order of a million million distinct cells. On the scale of size alone a large animal is as far removed from a single cell as the cell is from the molecules of which it is composed. But the cells are also grouped into patterns of great complexity, forming in the higher animals, at least, a great number of organs which carry out special functions. If a single bacterial cell is a marvellous construction, what can we say of an animal containing a million million of much larger cells, together forming a functioning whole?

This is a complexity of a totally different order from that of unicellular life and entirely new principles are involved in building such structures. Our knowledge of these principles is woefully inadequate but it is necessary to mention what is known about them because they are the basis of all the bewildering variety of forms of life which have developed during the course of evolution.

All such complex organisms begin as a single fertilised egg cell, which has been produced by the union of an egg cell (a rather large and complicated cell) produced by the female, with a sperm cell from the male. Both of these cells carry a single complete set of genes derived from the parents and as pointed out previously (p. 13), the fertilised egg cell now contains a double set of chromosomes, one from each parent. The fertilised egg cell now divides again and again and every cell so formed contains a double set of chromosomes. The result of the first ten divisions or so is a little ball (the blastula) of perhaps about a thousand cells, which all look similar, but after this, division continues and the blastula undergoes complicated changes as a result of which different parts of it develop into different tissues of the complete animal. Some cells develop into the digestive system, some form liver, others produce the skeleton, the skin or the nervous system. This remarkable process continues until an embryo is formed—in many respects a miniature individual which will grow until it is capable of an independent life.

Little is known as to how this process—perhaps the most extraordinary of all natural phenomena—is controlled and guided. Yet controls there must be, and because of the complex outcome they must be of the utmost delicacy, yet they act with great precision. The ability to develop such controls was a necessary precondition for the evolution of complex organisms and was a crucial step in the ascent of life.

Differentiation

There are clearly two stages in this process: first the formation in the growing blastula of *differentiated* cells, each kind of which has a different destiny; secondly the development from such differentiated cells of the functioning tissues of the complete plant or animal. Very little is known about the mechanisms which guide either of these processes, although the morphological changes which can be seen under the microscope have been described in great detail. The crucial process is perhaps the first. At some stage in the process of cell division, cells with distinct characters and destinies are formed. This process is known as differentiation. Little is known as to how it happens. It has been realised in recent years that the original egg cell itself is a highly complex object and that its cytoplasm (i.e. the parts outside the nucleus) contains in different regions substances which are capable of guiding the development of parts of the developing blastula in different directions. Thus it seems that while the male sperm provides only the genes of the male parent, the female egg also contains, besides the genes of the female parent, factors which guide the organism through at least the early stages of development.

Dr. J.B. Gurdon of Oxford University has carried out some remarkable experiments which support this. They are based on inserting nuclei taken from already differentiated cells of various tissues into the eggs of frogs and toads, which have been deprived of their own nuclei. In a considerable number of cases it was found that the egg cell with the transplanted nucleus developed into a complete individual. For example, when nuclei obtained from intestinal cells of tadpoles of the toad, were transplanted into enucleated egg cells of the same organism, complete swimming tadpoles developed. Thus a nucleus which had already become part of the cells of a differentiated tissue was capable, when placed in the egg cell, of starting again and going through the whole process of development, which is evidently controlled not by the nucleus itself but by substances present in the egg cell.

The amount of DNA in mammalian cells is much greater than that in bacteria. For example, the cells of the higher mammals contain enough DNA to form several million genes, i.e. about a thousand times as much as in most bacterial cells. Even allowing for the possibility that multiple copies of some genes occur, this is far more than required, on any reasonable estimate, to code for all the proteins of the animal. It is likely that many of these genes have regulatory functions which are not understood at present, but even so there is enough DNA to provide the different requirements of the many different types of cell present in the whole organism.

The crucial problem of the formation of complex organisms is how many distinct types of cells can be derived from an original cell which contains *all* the genes of the organism. One method would be for the various genes required in particular cells to be sorted out and segregated in the cells in which they are required to function.

This was not the plan which developed in the course of evolution, possibly because the sorting out the genes in this way presented too many difficulties. It is found that *all* the genes are present in every cell of the organism (with a few exceptions), but only those genes which are required to function in each type of cell become active—the remainder are 'repressed'. For example, a liver cell produces a different set of enzymes from those produced by a cell of the stomach lining or a cell which produces hair. All these cells have the same full complement of genes but only a selection of these genes, i.e. those which are required to give the cell its special character, are active in each type. It follows that every cell in a complex organism must have many genes which are inactive and indeed may never function throughout the life of the cell. For example the genes which are required only for liver function are permanently inactive (repressed) in kidney cells.

The basic behaviour of complex organisms is the ability to enable a selection of genes, and these only, to function in each type of cell. It was this ability which made complex organisms possible.

How is it achieved? What determines whether a gene is active or inactive in a particular type of tissue? We only have some hints as to the kind of mechanism involved. Some guidance on this question was obtained by Professors Francois Jacob and Jacques Monod of the Pasteur Institute, Paris, in their experiments on the simpler case of gene control of bacteria. Here it is also found that genes can be either active or inactive (repressed). In bacteria many genes are only active if they can be usefully employed.

Thus, as a rule, only those genes are active which produce the enzymes required to deal with the nutrient substances which are actually present in the medium in which they are growing, while those which code for enzymes which are not needed are 'repressed'. Monod and Jacob found that when a gene is required to become active, it is activated (derepressed) by an 'inducer' substance. When a new nutrient substance is present in the medium in which bacteria are growing, those genes which produce the enzymes which are required for its utilisation are activated, so that in effect nutrient substances induce the formation of the enzymes which are required to deal with them.

This example indicates that gene control mechanisms which can be quite complex occur even in bacteria. It is possible that similar mechanisms also occur in complex organisms, but very little is known of the processes whereby genes are controlled in such organisms, yet the very existence of such organisms depends on their having controls which function accurately throughout life.

Some very interesting experiments made by Professor Henry Harris, at Oxford University, have shown that inducer substances are in fact present in the cellular contents of animal cells, which are capable of activating genes which are normally repressed.

Professor Harris's experiments involved the making of hybrid cells from two tissues which might even come from different animal species. Two unlike cells are induced to combine by the presence of a virus which has been made innocuous by exposure to ultraviolet light. Probably the effect of the virus is merely to dissolve the cell wall at the point of contact so that two cells can fuse together forming a hybrid cell which contains the genes of both types. Hybrid cells have been made in this way for example from cultures of mouse or rat and human cell types, and also involving various types of chick cells. They retain many of the biochemical abilities of the original cells, e.g. they can, at least for a time, synthesise both DNA and RNA, and can even undergo a few cell divisions, but usually the hybrids do not survive very long.

Using this technique Harris carried out some profoundly significant experiments with the red blood cells of chicken. These contain the genes of chicken, but they are almost completely inactive, as these cells synthesise no protein and little RNA. Hybrids were made of these cells with a type of mouse cell, the result being a mouse cell also containing genes from the chicken cells. It was found that the chicken genes, which were almost completely inactive when they were in the chicken red cells, now became capable of the synthesis of the RNA. The chicken genes, which are almost completely repressed in the chick red cell, had evidently become activated when placed in the mouse cell, so it was concluded that the latter contained substances which were capable of activating (i.e. derepressing) the chick genes.

These experiments give the most conclusive evidence so far that the cells of complex organisms do contain activators or inducers of gene activity and that the behaviour of the genes depends on the nature of the cellular contents to which they are exposed. In order to understand what determines the gene activity in any type of cell we should then ask how it comes about that suitable inducers are present at the right time in every type of cell. If the inducers are proteins they will also be the result of gene action; if not, they will be formed by the action of enzymes which are themselves the product of gene action. At some stage in the development of differentiated cells the genes which give rise to the appropriate inducers must themselves have been activated and we should then have to enquire how this happened. We see that there must have been a succession of events perhaps going back to the very early stages of the development of the organism and possibly even to the egg cell itself, which, as we have seen, already shows indications that its various parts are to develop into different parts of the organism.

The picture we arrive at is that at the beginning, in the unfertilised egg, all the genes are repressed. After fertilisation gene activity begins and a step by step derepression of genes occurs, leading to differentiated cells, each with its own pattern of gene activity. It is evident that there must be in the developing embryo a complicated programme of events which leads eventually to the differentiated tissues. Little is known about the nature of the programme and how it is controlled.

It will be evident that notwithstanding the great progress which has been made in understanding the modes of operation of micro-organisms, the fundamental principle which enables a complex organism consisting of many types of cells to develop from a single fertilised egg cell, has not yet been found, although approaches to this problem are being made especially by J.B. Gurdon and Henry Harris.

Factors controlling growth

When all the necessary differentiated cells have been formed, they must multiply by cell division so as to form the various tissues and organs of the organism, while doing this they must retain their distinctive character. The different tissues must develop harmoniously in accordance with a predetermined pattern so as to produce the characteristic features of the new organism. It is well known that these features are inherited. In addition to the major features which distinguish, for example, a cat from a dog, even minor features such as the shape of the head, the pattern of hair, in race horses, the ability to run fast and, in cows, the yield of milk, are inherited to some extent. Little is known as to how such characteristics are transmitted. They are obviously gene controlled but it is not known how the genes influence and control the pattern of growth.

Nevertheless substances which control the development of certain organs, were discovered in the early part of the twentieth century, when it was found that various glands in the body produce substances called hormones, which exert powerful effects on other bodily tissues. Thus, it was discovered that the sexual glands produce 'sex hormones' which cause the development of the secondary sex characteristics and govern, in females, the sexual cycles. Over a hundred years ago, Berthoud showed that the sex glands of a cock, transplanted into the hen, caused the hen to develop a cock's comb. The capon (incomplete male) develops only a rudimentary comb but the injection of an extract from the male sex glands of any related species causes its full development. The genes for the cock's comb are evidently present even in the females but they require the stimulus of the male sex hormones before they become active.

Hormones derived from the ovaries of females also cause the development of female characteristics and cause females to come on heat. A large number of such substances have been isolated, both natural and synthetic, which bring about a great variety of effects.

In some cases the hormone-producing glands are themselves controlled by another gland which produces hormones which stimulate them. For example, the pituitary, sometimes known as the hypophysis, a small gland present at the base of the skull, produces a great variety of secretions which stimulate other glands including the sex glands to produce their own hormones. It also produces a 'growth hormone', which regulates the growth of young animals to adult size. A deficiency of this hormone causes stunting, while an excess leads to gigantism. This gland also produces prolactin, a hormone which causes the development of the mammary gland and stimulates milk secretion.

Although a considerable number of hormones having specific effects on particular organs are known, substances which control the size and shape of particular tissues have not generally been discovered, but they may well exist. To some extent the size of organs is self-regulating—they become large enough to deal with the demands made upon them—but there remain many features of all organisms which must be gene controlled.

It will be evident that much remains to be learnt about the influences which determine the size and shape of organisms. Yet we cannot conclude that there must be some special vital force in every organism which determines its shape and functions. The possibilities of scientific explanation have not been exhausted, although it may take many years before a complete and adequate explanation is achieved. We must expect that as the study of the functioning of living organisms proceeds, the complex interaction between the parts which leads to harmonious growth will eventually be understood and the way in which genes exert their control over the final result will be elucidated. It is a long step from the evolution of particular proteins to the evolution of structural features like the long neck of the giraffe or the shapes and patterns of butterfly wings. Inherited variations of such features must axiomatically be due to gene changes, but the nature of the genes concerned and the way in which they operate are almost completely unknown. From the study of simpler cases such as bacteria and from studies of chromosomes such as those of fruit flies and because DNA is a universal constituent of chromosomes, it is inferred that the genes which control structural features are also composed of DNA and it is to be expected that as in the simpler cases the DNA carries codes for proteins (indeed no other function for DNA is known) but what these proteins are and how they act has not been discovered. It may be suspected that they are similar in nature to hormones. If the genes which control structural features of organisms are composed of DNA, as we must expect them to be, then evolution will have been due to changes in the DNA complement of organisms. As we have seen, changes in DNA can be produced by errors of the replication processes which may result sometimes in the duplication of genes or occasionally in their deletion, by the action of radiations such as cosmic rays, X-rays or ultraviolet light by the capture of genes from other organisms and by the effects of certain chemicals, which may occur in the environment. These agents generally act on DNA in a haphazard way and which gene is affected and in what manner is a matter of chance. It has therefore been suggested by Jacques Monod that the variations which have given rise to evolution are controlled solely by chance since all sorts of mutations may occur and the only control is the negative control of the environment which eliminates those which are unfavourable.

In my opinion this conclusion is far from certain because of the great complexity of the situation and the probability that changes of structure are produced by a very complex interplay of effects, the nature of which are largely unknown, although basically they must be determined by the genes directly or indirectly. It may well be that mutations occur more readily in some directions than in others and that some types of change occur more frequently than others, e.g. there may have been a continuing tendency for the neck of the giraffe to lengthen owing to the ways in which the genes concerned interact with each other.

The critical factor in such cases might be the loss of genes which inhibit the growth of the neck and these genes may be susceptible to change or loss more easily than others. This is not the place to pursue such difficult technical questions, but it will be clear that much remains to be learnt, indeed the first principles are hardly known, as to how life has evolved the fantastic variety of species which are alive today and which having appeared during the course of evolution, flourished for a time and then failed to meet the test of survival in a competitive world.

Mutations brought about artificially are nearly always deleterious and cause a loss of some structural feature or function.* We do not know how

---

*This is not quite accurate as some useful mutations have been brought about by exposing plants to nuclear radiations. Of course such mutations will not necessarily have survival value in the wild.

to bring about mutations which are advantageous, or seem to be so from our point of view. In fact the only available method is the well tried one of the animal or plant breeder—to take advantage of variations which have occurred spontaneously or to hybridise varieties so as to produce strains which have a desirable combination of characteristics. The possibilities of genetic engineering, i.e. of putting together combinations of genes into an organism which might be thought to be useful or of making new genes artificially are largely illusory with our present knowledge of genes and how they operate.

The much discussed question of 'test-tube babies' may be mentioned here. A group of Cambridge medical scientists headed by Dr. E.S. Edwards have successfully removed human ova from a mother's ovaries and fertilised them with a donor's spermatozoa outside the body. It has been stated that fertilised ova have been successfully implanted into the mother's body and have led to a successful pregnancy. This very difficult process must be regarded as a surgical intervention to permit a woman with a gynaecological defect such as blocked Fallopian tubes, down which the ova pass prior to fertilisation, to bear children and does not carry any more ethical implications than e.g. artificial insemination (AID) or birth by Caesarian section. The further problems of keeping the fertilised ovum alive and developing normally outside the body has not been solved to a more than limited extent. Dr. Denis New and associates at Cambridge have succeeded in keeping rat embryos of various ages alive in culture media outside the body for a day or two.

Science fiction writers have also seized on the experiments of Dr. Gurdon in which cells of body tissue are introduced into egg cells of the same species, which then develop into complete individuals, to suggest that it will be possible to produce a large number of replicas of a human being by persuading his body cells to dedifferentiate and develop afresh into a new individual. It is suggested that this might even be possible with cells maintained in a frozen state long after the person from whom they were taken is dead. Certainly it is believed that all the cells in the body contain the complete genetic information for the whole individual, but even if it were possible to induce the nucleus of a body cell to enter a female egg cell and to grow into a new individual, the product would still be uncertain as we do not know which genes would be dominant, the mother's or the donor's and in any case it seems that many of the factors which control development, which must be of paramount importance in the formation of a new individual, are in the egg. Also abnormalities of development easily occur even in the natural state and result in defective individuals. Thus it need hardly be said that such fantasies far outrun the actual state of knowledge now or in any reasonably anticipated future.

For example, normal human females have two X chromosomes, while males have one X and a small chromosome called Y. Occasionally individuals with an abnormal complement of chromosomes such as XYY are produced. It has been found that this combination is associated with criminal tendencies. Another condition, known as Klinefelter's syndrome, is due to the combination XXY, in which the normal maleness produced by XY is modified by the additional X chromosome, giving female character- istics. In other cases (Turner's syndrome) one X chromosome has been lost and this results in an incomplete development of female characteristics. Another chromosome abnormality concerns chromosome number 21 (possibly 22) of which two are normally present. In about 1 in 1000 births three of these chromosomes are present and the result is a condition known as mongoloid idiocy. This results from an abnormality of the mother's egg, which contains two instead of one of these chromosomes and this abnormality increases markedly if the mother is aged over forty. These and many other conditions involving chromosome abnormalities shows that the chromosome complement of one or other of the germ cells is easily upset and the drastic treatments envisaged by the proponents of 'genetic engineering', even if they survived so far, would be likely to produce a greatly increased crop of abnormal births.

8 Medical applications of biology

The great advances of biology and especially biochemistry during the present century have revolutionised medicine in some directions, although others remain obdurate. Before discussing other aspects of life, in particular how the behaviour of the higher animals is organised, I shall give a short review of some of these developments.

Antibiotics

It is difficult to think of a more perfect breeding ground for parasitic bacteria and viruses than animal cells and fluids and it is perhaps not surprising to discover that many forms of life have developed devices which give some protection against such agents.

One such device is the elaboration of substances which interfere with an important and necessary feature of the life of the invading organism. The first of these 'antibiotics', penicillin, discovered by Alexander Fleming in 1929, was isolated and prepared on a usable scale by Howard Florey and Ernst Chain in Oxford during World War II. The product of the common fungus *Penicillium notatum,* it interferes with the life of the type of bacteria known as Gram-positive (because they are stained by a dye invented by the bacteriologist Gram). The effect of penicillin is to prevent the formation of a characteristic substance from which their external coat is made. Since penicillin has no effect on animal cells, which do not have this type of substance in their cell walls, it can be given in large doses and had a remarkable success in controlling numerous bacterial infections and diseases.

Numerous other effective antibiotics were soon discovered. In 1940 Waksman obtained a substance called *actinomycin,* from a strain of *Actinomyces,* which interferes with the synthesis of messenger RNA. Because this process is essential to animal life, actinomycin is also toxic to animals and cannot be used in medicine except in some special cases. Waksman also isolated *streptomycin* from a soil organism; this is a substance which is effective against many Gram-negative organisms, e.g. those responsible for tuberculosis, meningitis, pneumonia, *Shigella* dysentry. It is rather toxic because it interferes with ribosome function, but is a useful supplement to penicillin in some disorders.

Another group of useful compounds are the tetracyclines, which interfere with protein synthesis, and chloramphenicol, which can be used against Rocky mountain fever, undulant fever and typhus.

These compounds act as 'antagonists' of some aspect of the life process, usually because they are very similar to substances which have essential functions in the normal processes of life. The first examples of such antagonisms to be discovered were the *sulphonamides*—synthetic substances which were found to be effective against many bacteria. It was discovered by Fildes and Wood that the bactericidal action of these drugs was interfered with by a similar compound PABA (*para-aminobenzoic acid*) and they concluded that sulphonamides and PABA are antagonistic to each other. PABA is a constituent of important vitamin-like substances, such as folic acid, which are essential to the life of many bacteria and therefore in the absence of PABA or when it is replaced by an 'antagonist' the bacteria cannot grow. The antagonist acts because it is sufficiently like PABA to take its place in some reactions, but not close enough to it to act perfectly. This is the principle on which many of the antibiotics mentioned above bring about their effects.

It may be asked why, if antibiotics interfere with some essential aspect of the life process, they do not interfere equally with the host organism. Obviously to be useful there must be a differential effect between the action on the bacterium and its host and this exists in many cases to a greater or lesser extent. Bacteria are often vulnerable because their biochemical equipment has been reduced to the lowest extent which still permits life to continue. If a single one of the essential processes is interfered with, the bacterium cannot live, or at least cannot multiply. On the other hand in animal cells there is a greater profusion of processes available, and often if one process is interfered with, the cells can find an alternative pathway to produce the necessary substance, so that in general animal cells are less vulnerable to antibiotics, especially to those which produce a highly specific effect.

The use of antibiotics has led to the control and in some cases the effective disappearance of many contagious diseases. This has been a factor in the formidable rise of population in many parts of the world—especially among the dense populations of the tropics. Another important factor has been the reduction in number of insects, the primary carriers of diseases like malaria, due to the development of effective insecticides such as DDT during World War II.

The widespread use of penicillin and other antibiotics has led to the appearance of resistant strains of bacteria (see p 32). Certain individual bacteria are capable of producing an enzyme, penicillinase, which can destroy penicillin. Normally the presence of a few bacteria having this property would hardly be noticed, but the destruction of bacteria which are sensitive to penicillin gives the resistant strains a chance to multiply (since resistance is inherited) and they have done so to an alarming extent. Indeed one hospital in recent years faced with a great increase in infections found that the situation was improved if they stopped using antibiotics completely for a time!*

It was also found that giving antibiotics like penicillin and streptomycin to farm animals increased the rate at which they grew—presumably because the antibiotic produced changes in the bacterial population of the animals' guts which had a favourable effect on their growth. Animal foods were put on the market which included suitable antibiotics. This procedure, however, had unfortunate effects in other directions. It obviously produced in the animals themselves a great increase in the population of resistant bacteria and since the people who handled the feeding stuffs were obviously also continually dosed with the antibiotic, they were unlikely to respond very much if the time came when they needed the antibiotic. The result of an enquiry held in London under Sir Michael Swann in 1966 stimulated the government to issue a regulation forbidding the practice of adding antibiotics to animal foods, except for certain substances which are unlikely ever to be used in medicine.

Antibodies and immunity

One of the most remarkable inventions of the animal body is the immune response which it has developed against foreign substances and foreign organisms. The existence of some kind of protective mechanism is indicated by facts which have been known from earliest times, for instance it is well known that individuals who have recovered from infectious diseases like mumps and measles are usually immune to a second infection by the same organism.

It was realized in the eighteenth century that dairymaids who had contracted cowpox, a mild form of smallpox, were immune to the latter and Edward Jenner a doctor in Gloucestershire, confirmed this by injecting matter from cowpox sores into healthy individuals and found that it conferred immunity against the dreaded smallpox. This was the origin of vaccination against the smallpox, which has led to the virtual disappearance of the disease in most countries.

In consequence numerous attempts were made to discover mild types of diseases which could be used to confer immunity against the more virulent types. Pasteur, for example, found that very old cultures of chicken cholera produced a mild form of the disease which conferred immunity from the virulent form of the disease.

A successful result was rarely achieved in this way and attempts were made to produce attenuated forms of organisms by other means. It was found that in some cases, even when the organism was killed by heating so that it was no longer infectious, it was still capable of conferring

*It is not clear why this should be so—perhaps bacteria compete among themselves and a mixed population is less harmful than a single one.

immunity. Similarly, organisms which have been inactivated by treatment with chemicals may still be capable of giving rise to immunity. This is also the case with diseases caused by viruses. For example the Salk polio vaccine, which was obtained by treating polio virus with formalin. This makes the virus incapable of multiplying, and the vaccine confers immunity against polio and has led to a virtual disappearance of this disease in many countries.

The existence of such immunity led to investigations of its mechanism and it was found that the blood of immune persons contained substances which caused the agglutination (clumping together into sizable particles) or precipitation of the bacterium (or virus) to which the person was immune. The agglutinated organisms were often more easily attacked and digested by the white cells (phagocytes) in the blood, which act as scavengers. The substances in the blood which produce these effects were called *antibodies*.

It was found that the whole bacterium or virus was not required to produce this kind of response. Parts of bacteria such as cell walls were sufficient and even simpler substances such as proteins and some carbohydrates could provoke an immune response. Substances which induce the formation of antibodies are called *antigens*. In experiments with such substances, it was found that antibodies were highly specific in their action. An antibody which precipitates one antigen will not precipitate another which may be only slightly different. Numerous experiments were carried out by Karl Landsteiner from 1929 onwards, to see how far he could modify a protein by chemical processes without destroying the ability of the original antibody to precipitate it. He found that frequently only a slight modification of the protein was sufficient to prevent its precipitation by the antibody. Similar proteins from different species e.g. egg white from hen and duck, also behaved differently and the antibody to the first would not recognise the second.

Antibodies which have been formed in response to infections may remain in the blood for considerable periods—sometimes during the whole life of the individual. The fraction of the blood protein called γ-globulin was found to be made up almost entirely of the various antibodies a person possesses. These antibodies can be used to confer immunity (e.g. against measles) on another person by transfusion of the blood or the part of the blood which contains the antibodies. This, like blood transfusion in general, is a somewhat dangerous practice as there is a risk of transferring unwanted substances and possibly virus particles at the same time. It is now recognised that the virus of infection jaundice can be transferred by blood transfusion.

Biochemists and pathologists were naturally interested in the mechanism by which the appropriate antibodies were made. Antibiotics were found to be proteins of a special character. The first idea was that there is a process whereby the antigen itself guides the production of a suitable protein which will act as its antibody, i.e. the antibody is tailored to fit the antigen it is intended to combine with. This theory met with a difficulty which has proved to be insuperable. How does the antibody forming mechanism know which substances are foreign to the organism and which native to it? If antibodies were made for the proteins normally present, for example, the result would be disastrous to the life of the organism.

This difficulty was overcome by another theory, a *selection* theory, which was originally suggested by Professor Jerne of Pittsburgh in 1955 and developed by Dr. Talmage (Colorado) and Dr. (now Sir) Macfarlane Burnet (Melbourne) in 1957. This theory suggested that the ability to make a large number of different antibodies is present in the organism all the time. It was shown that antibodies are synthesised in white cells, known as lymphocytes, present in the blood. These are formed mainly in the bone marrow and circulate round the body in the blood and also to some extent in the 'lymphatic system' which connects the bone marrow with lymph nodes in which they are stored and also in the spleen and thymus. It has been found that each lymphocyte can produce only one kind of antibody. The effect of introducing an antigen is to stimulate the formation of those lymphocytes which can produce a suitable antibody by stimulating their cell division.

It is suggested that there are no lymphocytes producing antibodies for the substances normally present in the organism because any such lymphocytes have been swamped by self-proteins at an early stage of the animal's existence.

This system of producing immunity against foreign substances only begins to function shortly after the birth of the animal and any antigenic substance which is present before this time is accepted as normal. Thus, if an antigen is introduced into the newborn animal, it will throughout its life accept this substance as belonging to itself.

Many of the details of this remarkable process are still unknown, but it has been found that the thymus gland—a small gland present in the neck—is involved in the process of bringing the immune mechanism into action. It give rise to 'competence' i.e. it enables lymphocytes to be activated so that when they recognise the presence of an antigen to which they can respond they will multiply. The thymus thus acts as a censor of the numerous kinds of lymphocytes which the organism produces. Those which could produce antibodies against self-antigens are eliminated while those which could be active against foreign antigens are permitted to multiply.

How does it come about that such a variety of lymphocytes are available? Since antibodies are proteins, a lymphocyte which can make a particular antibody must carry in its DNA the code for this protein. Each type of lymphocyte can make only one antibody—as has been shown in experiments in which the antibodies produced by individual lymphocyte cells are studied. It follows that in the whole population of lymphocytes the DNA must carry a great many different antibody codes.

There has been a great deal of discussion as to how this comes about. It is generally believed at present that there is some 'diversifying' mechanism at work in the formation of lymphocytes from their primitive stem cells in the marrow which modifies the codes carried by the DNA in those genes which are going to code for antibodies so that a great variety of such codes are present, which are available to the organism if required.

Transplantation of tissues

It has been known for a long time that although it is possible to transplant small areas of skin from one part to another in the same individual, a graft almost invariably fails if it is attempted between different individuals. The transplant which at first appears to be healthy, eventually shrivels up and is rejected. An exception is the case of identical twins i.e. individuals of exactly the same genetic constitution, which will accept transplants from each other. A similar state of affairs was found in animals, in experiments by Sir Peter Medawar and his co-workers in London during the early 1960's. Colonies of rats with identical genetic constitutions can be obtained by inbreeding, and it was found that transplants with individual animals belonging to other strains were rejected. Whether a transplant would be accepted or not was evidently genetically controlled. Since with human beings and wild animals practically every individual (identical twins excluded) is genetically different, this confirms that transplants between one individual and another will not generally be accepted.

However, as in the case with simple antigens, if a graft is made at or soon after birth, the animal will accept other grafts from the same individual throughout its life. It thus accepts as part of itself any tissue present in it at birth.

It has been demonstrated that this kind of immunity is brought about not by antibodies but by lymphocytes themselves. When a foreign graft is made lymphocytes accumulate around the grafted tissue and cause it to die. That the presence of foreign tissue induces the formation of a highly specific type of hostile lymphocyte is shown by the fact that if a second graft of the same tissue is attempted after the first has been rejected, the rejection process occurs immediately because the appropriate lymphocytes are already present.

As with simple antigens, the thymus gland is necessary before this type of immunity can come into action. Dr. J.F.A.P. Miller showed that if the thymus is removed at or just before birth the rejection of foreign tissues did not occur—although without a thymus the animals eventually died of a wasting disease.

The phenomenon of tissue incompatibility provides a great barrier to the replacement of damaged organs by surgical transfer. No-one can live very long with totally impaired kidney function, yet life continues quite comfortably for long periods with a single functioning kidney. This leads to the idea of transplanting kidney from a healthy individual to a person with non-functioning kidneys. This is a transplantation with the same restrictions as skin transplants. Between identical twins there is no difficulty, but between any two non-identical persons rejection of the transplanted kidney may occur after varying periods. The period can be lengthened by putting the immune mechanism of the person receiving the transplant out of action. This can be done by the action of X-rays on the bone marrow or by treatment with certain 'immunosuppressive drugs' of which azothioprine (Imuran) and 6-mercaptopurine are the most commonly used. These drugs prevent the formation of new lymphocytes by cell division, so that the anti-graft reaction cannot take place. The immunosuppressive treatment has to be continued for a long time, even perhaps indefinitely, and during this period the person has no protection against invasion by bacteria and other organisms. Another immuno-suppressive treatment is the use of anti-lymphocyte serum (ALS). In this procedure, lymphocytes from one animal are injected into another, not necessarily of the same species, and since they are foreign to the recipient, antibodies to the lymphocytes are formed. It was found that these antibodies had a powerful 'immunosuppressive' effect i.e. the recipient's ability to make other antibodies was greatly impaired. This serum (ALS) can be used in transplant operations, but its use has some disadvantages.

Transplanting sound organs from recently dead persons is also being widely practised. The degree of incompatibility varies considerably and the time taken to reject a transplant is accordingly variable. Methods of 'tissue typing' have been developed which make it possible to estimate the degree of incompatibility between two individuals so that suitable donors can be selected in some cases.

Blood groups

It has been known for a long time that blood transfusion from one person to another is sometimes possible and sometimes disastrous, because in some cases the mixture of the two bloods causes the red cells to agglutinate. The examination of different bloods has led to their classification into 4 main groups terms A, B, AB and O which are defined as follows: (1) blood of group A is agglutinated by serum from group B; (2) blood of group B is agglutinated by serum from group A; (3) blood of group AB is agglutinated by sera of both groups A and B; and (4) if the blood is not agglutinated by either A or B serum it is group O.

An explanation of this behaviour was found by Landsteiner in 1900. Blood of types A and B have red corpuscles which carry characteristic antigens A and B. The sera of the various bloods contain antibodies which cause agglutination when the respective antigen is present. An antibody to type A blood is not usually present in type A blood serum as the blood would agglutinate itself if it were present, but an antibody to type A (anti-A) is present in type B serum and type 0 serum, both of which can agglutinate type A blood. Similarly an antibody to type B corpuscles (anti-B) is present both in type A and type 0 serum. Type 0 blood corpuscles do not contain any antigens which react with the antibodies present in other types of serum, hence it is not agglutinated by sera of either A or B. This situation is summed up as follows:

Blood type	0	A	B	AB
Antigens	none	A	B	A and B
Antibodies	anti-A, anti-B	anti-B	anti-A	none

Since antigens are produced by the action of genes, blood groups are inherited in the same way as other genetic characteristics.

An understanding of this remarkable system has saved countless lives as with its aid it is possible to make blood transfusions with a certainty that a blood which is compatible with that of the patient has been chosen.

Hormone Therapy

Another class of human diseases is due to the malfunctioning or atrophy of the glands which produce various hormones. Since the hormones of animals are the same or similar to those of human beings, these conditions can often be corrected by injection of the same hormone from an animal source. For example, it was found that diabetes, which is due to the inability of the organism to utilize carbohydrates, with the result that sugars accumulate in the blood, could be produced artificially in animals by the removal of the pancreas and could be alleviated by injecting pancreas extracts. In 1921 Banting and Best at Montreal isolated insulin from pancreas extracts of animals, a protein which when injected alleviated diabetic symptoms. In many cases diabetic patients have been maintained in compariative comfort for many years by regular injections of animal insulin.

The desire of older men to have their virility restored by injection of extracts of monkey gland proved to be unattainable. But nevertheless gland extracts or their chemical equivalents, which the drug houses have produced in a great variety, are widely used to produce specific effects. For example, cortisone, a product of the adrenal glands, alleviates rheumatism and inflammatory conditions.

The sexual organs of both males and females require for their correct functioning a variety of hormones, the synthesis of which is controlled by the pituitary, a gland situated at the base of the skull. The female sexual cycle is brought about by hormones which are synthesized in the ovary, viz. oestrogen which initiates the menstrual cycle and progesterone which controls the later stages. The formation of these hormones is stimulated by two hormones of the pituitary, LH and FSH*. It was found that the formation of these hormones in the pituitary was repressed by maintaining a concentration of oestrogen and progesterone in the blood throughout the menstrual cycle and if this is done, the menstrual cycle and the release of ova and implantation subsequent to fertilization do not take place. A mixture of an oestrogen and progesterone-like substance, used as a birth-control pill, which can be taken by the mouth, has had enormous social effects and is responsible more than any other factor, for the great changes of sexual morality which have occurred during recent years.

Another hormone, also a product of the adrenal glands, which has a large indirect effect on human life, is aldosterone, discovered by Drs. S.F. and S.A.S. Tait in London in 1957. This is the principal regulator of the sodium (salt) excretion by the kidneys. When, as happens with elderly men, the adrenals do not produce enough aldosterone, the effect is a rise in blood pressure in an effort to compensate. This 'hypertension' was one of the principle causes of strokes and other disabilities of the elderly and can be largely eliminated by treatment with aldosterone substitute and this has resulted in an increased life for many elderly people, one of the factors which has resulted in a considerable increase in the numbers of the population over 65.

Cancer

The diseases hitherto mentioned are due to the invasion of the body by extraneous 'parasitic' organisms such as bacteria or viruses. There are other conditions which are due to the malfunctioning of the body in some respects and these have proved to be much more difficult to deal with. The commonest and most serious of these is cancer, which is due to a cell or group of cells escaping from the normal controls, whatever they are, which maintain the body in its normal state with all its parts in proper proportion to each other. The abnormal cells first give rise to a 'growth' or tumour but at a later stage abnormal cells from the tumour may escape and be carried by the blood into other parts of the body, where further growths are established. Such tumours are called malignant.

It was at first thought that tumours were due to parasitic organisms, such as bacteria, but about 1930 following the observation that contact with tars led to skin tumours, scientists at the Royal Cancer Hospital, London, established that painting the skin of rodents with pure chemical compounds resulted in the formation of skin tumours. The original

*LH = luteinizing hormone; FSH = follicle stimulating hormone

compounds which produced this effect belonged to a class of hydro-carbons, present in tar, but it has since been found that malignant tumours in animals can be produced by contact with or injection of a large variety of diverse chemical compounds, some of a very simple character. Tumours are also produced by exposure to X-rays and radio-active radiations, e.g. those emitted in nuclear fission.

Besides simple chemical compounds it has been found that certain types of cancer, especially leukemia, a cancer of the blood forming cells, which results in a great excess of white blood corpuscles, can be transmitted from one animal to another by viruses. This has never been demonstrated to occur with human beings, because it is impossible to inject into a human being a preparation from a tumour which might contain a cancer virus. Nevertheless it is most unlikely that human beings are exempt from viruses which commonly occur in animals.

Tumour viruses may be based on either DNA or RNA. In the latter case Dr. H.M. Temin and D. Baltimore working in the Massachusetts Institute of Technology and the Salk Institute, La Jolla, respectively, in 1970 onwards, made the astonishing discovery that there is an enzyme in animal and other cells which convert the RNA virus into the equivalent DNA form, which is the reverse of the usual transcription process DNA→RNA (see p. 26). In the DNA form the virus may now be incorporated into the DNA of the chromosome, where it can remain during cell divisions. The possibility of easy transfer of DNA between bacteria and other cells leads to the alarming possibility that tumour viruses might find their way into human tissues after first being inserted into bacteria such as *E. Coli.,* which are commonly harboured by the human gut (see p. 31). A code of practice drawn up by groups of scientists both in U.S.A. and England recommends that such experiments should only be done with bacteria which cannot live in man.

During recent years great efforts have been made by scientists to determine how it is that simple compounds and other agents upset the normal life processes so drastically. It was discovered that most of these compounds, like X-rays, were 'mutagens' i.e. were capable of causing mutations to occur when fed to various animals such as fruit flies or mice (see p. 31). The reason for this was they, or compounds into which they were converted in the body, were capable of reacting with the DNA of the cells and so affecting the genes of the organism concerned.

Thus the theory that cancer was due to a genetic change of some of the body cells, which destroyed, in particular, some of the genes responsible for the control of growth, was put forward and accepted by many of the workers in this field. As a result of the damage to the genes, the cells concerned could multiply without the normal restraints, whatever they are, which keep tissues from growing abnormally, e.g. which ensure that a finger remains a finger. The result of abnormal growth of some cells is a tumour.

This theory has been widely accepted at least as a working hypothesis, although there are many features which are not understood. However the genes responsible for normal growth controls have not been identified, and it is not known how they bring about these effects. Also why do certain groups of cells when exposed to these agents develop into a tumour, but not others?

The understanding of these effects requires a deeper knowledge of the life processes than has been obtained so far. The theory also does not suggest possible means of curing cancer, or of preventing it, other than by avoiding exposure to the agents which have been shown to produce it in animals, e.g. tobacco smoke which contains cancer producing tars.

No way of bringing aberrant tumour cells back under control has been suggested or discovered and the methods of treatment used attempt to destroy the aberrant cells in various ways. These include:

Surgery: Since the tumour is initially a local condition, if it can be removed completely at an early stage a complete cure is effected.

X-rays: Exposure to suitable doses of X-rays or similar radiations can kill cells. (X-rays may also cause tumours to occur, but the incidence of this is rather low. The commonest effect is to kill the cell.) In this treatment it is difficult to ensure that every cell has been killed, hence the tumour may recur after X-ray treatment.

Drugs: Certain drugs exist which have a lethal effect on cells in the body. As it is usually impossible to apply them locally, they are injected and carried round the body in the bloodstream and these have a deleterious effect on the whole organism as well as on the cancer cells. The final result is thus a balance between lethal effects on tumour cells and the toxic effects on the body as a whole. Certain drugs are highly effective with certain types of cancer.

Immune processes: The tumour is in a sense a foreign organism and it has been suggested that the body could be stimulated to reject it in the same way as it rejects transplants of tissues which are not identical genetically. (See p. 6). Would it be possible to stimulate the defence mechanism e.g. by an anti-tumour vaccine, so that it will reject tumour growths? One difficulty is that tumours are very varied in character and it would be necessary to have a vaccine for each of the many types of cancer. Also the immune response to tumours, although it has been demonstrated, seems to be rather weak. Sir Macfarlane Burnet however thinks that spontaneous tumours start frequently in animals and humans, but they are usually destroyed before becoming established, by the normal immune processes, which keep up a continuous surveillance of the body for aberrant cells.*

It is evident that although progress has been made in understanding cancer, much remains to be learnt about it, and the methods of treatment

*In Macfarlane Burnet, *Self and Not-self*. Cambridge Univ. Press, 1969.

leave much to be desired. Notwithstanding the inherent difficulties mentioned above, progress has been made especially by using combinations of the various methods, and in the U.S.A. it is stated that more than one third of cases of cancer can be cured if detected at an early stage.**

A more promising objective will be to identify and eliminate as far as possible cancer producing agents from the environment.

**In 1973 President Nixon proposed an all-out programme to discover a cure for cancer, using the kind of co-operative effort which had been used so successfully to develop the atomic bomb in the early 1940's. The problem however, is quite different, since the atomic bomb required mainly the application of known scientific principles to a particular objective. In the case of cancer a much fuller knowledge of the control mechanisms of the body is necessary before the nature of the problem can be understood and this is best achieved by numerous investigators studying those aspects of life which interest them. No doubt some acceleration of this could be achieved, if more finance were made available; but there is no guarantee that even if the nature of cancer is well understood a cure will be possible, other than by the methods mentioned above.

9 The first glimmerings of mind

It is a great advantage to all organisms to be able to respond to the presence of useful substances, like oxygen or various nutrients, to the conditions of illumination or temperature or to the presence of moving objects. Devices for such purposes go very far back in evolution. Even unicellular organisms are sensitive in one way or another to a variety of outside influences. For example, some bacteria are not only sensitive to oxygen tension in the medium they are living in, but are able to make movements towards regions of higher oxygen tension. Certain strains of the well known bacterium, *E. coli*, possess flagella i.e. long filaments (see Fig. 15), which perform a wave-like motion by means of which the bacterium is propelled through the medium. Dr. Julius Adler of the University of Wisconsin has shown that this enables the bacteria to move towards regions of high oxygen tension, but the way in which the oxygen tension controls the movement is unknown. It could be that in the parts of the bacterium which are well supplied with oxygen or nutrients the flagellar motion is more vigorous, but it is not clear how the direction of the movement is controlled.

Other cells are sensitive to light. The congregation of a number of light-sensitive cells together on the surface of an organism produced light-sensitive spots which formed the first primitive kind of eye—which was only capable of distinguishing light and dark. In certain algae such sensitive spots are associated with flagella, which are stimulated when light falls on a sensitive spot and this provides a possibility of movement from darker to brighter regions or of orienting in the optimum direction.

The association of such light-sensitive spots with flagellar membranes led to a more organised structure, in which the light-sensitive cells are mounted on a stalk, forming primitive 'eyes' such as are found on the upper surface of the rays of star-fish.

Out of such primitive sensitivities the most extraordinary sense organs have developed during the course of evolution. They frequently operate to the limit which is physically possible. For example it has been shown by Dr. Pirenne and colleagues at Oxford that the human eye is capable of detecting a very few quanta of light—a quantum being the smallest physical unit of light. Birds especially have remarkable powers of vision; the great condor can see a small animal from the height of fifteen or twenty thousand feet.

Some animals, such as bloodhounds, have a remarkable sense of smell and can detect the smell of a particular human being for hours after he has gone by. Insects also have an extraordinary sense of smell for certain compounds. The sex attractant emitted by the female gypsy moth can be detected by the male at distances up to three miles. Ants and bees can distinguish by smell the members of their own colony from others. If a 'foreigner' stays in a hive or nest long enough to acquire its smell it is then accepted as a member of the colony.

The ability to detect particular compounds must be due to 'detector' cells which are sensitive to the compound and which respond to it. Other cells are sensitive to pressure and this has been developed in certain cells called Pacini corpuscles in which a slight distortion of the surrounding tissues produces a response of a kind to be described below. Many insects have an acute sense of touch due to sensitive hairs, the displacement of which distorts the sensory cells of the hair socket.

A similar sense has been highly developed in the higher animals to provide hearing. Sound is carried by vibrations in the air which produce an oscillation of pressure on the surfaces they encounter. In the human ear, an organ of extraordinary complexity, the sound causes a vibration of a membrane called the tympanum, which is transmitted by a delicate mechanism to a fibre of hair at the base of which is a sensitive cell which detects its movements—a device which shows a profound correlation with physical principles.

A sense organ by itself is useless; to be useful its responses have to be capable of influencing the behaviour of the organism it belongs to in some way. As I have suggested above, the primitive sense cells were associated with membranes which were capable of transmitting the response to attached structures like the flagella mentioned above. When the sensitive part of a cell is stimulated it is natural that the response will affect the cell membrane in its vicinity and this response may then easily be transmitted to neighbouring cells. Out of such responses developed the nervous system—with momentous effects on the development of life.

The sensitive cells themselves developed long tubular protuberances, a continuation of the cellular membranes, which in the higher animals may attain a very great length, through which the sensory response is transmitted to distant parts of the organism.

When such 'nerves' have been developed, the original excitation, set up in the membrane associated with the sensitive elements, is channelled down the nerve tubes and retains its character throughout the length of the tube. Much work has been carried out to discover the nature of the nerve impulse which is stimulated by the excitation of the sensitive elements at the nerve ends. A great deal has been discovered about it, but it is too technical to be described here.

The experiments were greatly helped by the discovery of the 'giant' nerves of the squid which may attain a size of 0.5 to 1 mm in diameter, and Professors A.L. Hodgkin and A.F. Huxley at University College,

London, used these fibres to investigate the changes which occur when a nerve impulse passes down the fibre.

The main characteristic is a breakdown of the electrical characteristics of the nerve membrane when the impulse passes by. Normally the solution in the interior of the nerve is well insulated from the exterior medium by the membrane itself, which consists mainly of a fatty substance. When the impulse arrives, this insulation breaks down and electrical charges are able to cross the membrane. This breakdown is only temporary and when the nerve impulse has passed, the properties of the membrane are restored to a state in which it is ready to receive another impulse. An impulse of this kind has something of the character of a pulse of an electric current, but it must be emphasised that the analogy is rather superficial. These impulses move down the nerves very much more slowly than electric currents move in metallic wires. The speed varies according to the size of the nerve. In the giant squid fibres it is about 20 metres per second; in smaller fibres less.

The nerve impulse continues along the nerve until it arrives at its termination. This may be another nerve cell—which the nerve does not enter but merely makes contact with across a membrane (called a synapse)—if the impulse is strong enough it may stimulate the second nerve cell to send an impulse down its own nerve fibres. It thus acts as a kind of relay station to transmit the original impulse. Alternatively in the case of 'motor nerves' the nerve may terminate at a 'locomotor' organ, i.e. a large group of cells (muscles) which are capable of contracting. The arrival of a suitable nerve impulse at the nerve end causes a liberation of a substance called acetylcholine at the points of contact of the nerve with the muscle membrane and this causes the muscle to contract.

When the newly invented electron microscope was applied to the study of thin sections of muscles, a very remarkable arrangement was seen by Dr. H.E. Huxley at King's College, London in 1954, from which he and Professor A.F. Huxley, Drs. Jean Hanson and R. Niedergerke developed the current theory of muscle action. The muscle fibre in fact consists of two kinds of rods or filaments, one of which is fixed and the other movable. When muscular contraction occurs the movable filaments slide inside the stationary rods so as to shorten the length of the muscle fibre as a whole (Figs. 20, 21).

When they contract in this way the muscle fibres are able to exert a considerable force on the tendons to which they are attached. Although a great deal of work has been done on this mechanism some features are still awaiting clarification. In the electron micrographs tiny 'hairs' can be seen between the mobile filament and the stationary rods and it has been suggested that those hairs, which belong to the moving filament, first attach themselves to the stationary rods and then undergo a shortening, which pulls the movable filaments along. The arrangement may be likened to a rack and pinion arrangement in which cogs attached to a rod cause it to move along a stationary 'rack'. Movements against the forces

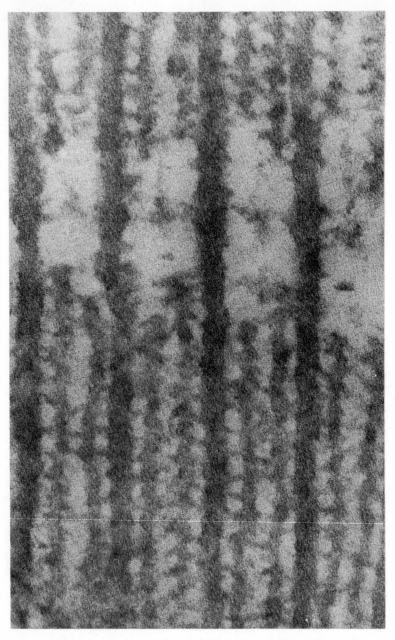

Fig. 20   Electronmicrograph of a section of muscle. (By Dr. H.E.
          Huxley.)

acting on the muscle requires energy and there is no doubt that fuel is used up when muscles operate which is provided by the oxidation of sugar, but the way in which the chemical processes providing the energy are coupled with the contractions causing movements is still unknown.

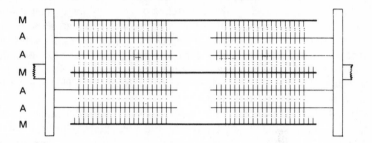

Fig. 21   An element of striated muscle as suggested by Huxley and Hanson. The actin fibres (A) are moved along the myosin fibres (M) by means of a kind of rack and pinion mechanism. The traction is exerted by fine filaments of protein indicated by double dots, which connect the two kinds of sliding fibre.

The whole mechanism, clearly one of the most extraordinary of life's inventions, is a remarkably efficient method of converting chemical energy into work. Movement of the muscle fibres is started by impulses arriving down a 'motor' nerve. It is therefore under the control of the animal's nervous system, which is actuated by the information received from the sense organs in ways we will discuss in the following chapters, so that the animal is able to make movements which are indicated as desirable by the information given by the senses.

The connection between the sense organ and the motor apparatus however is not direct except in the very simplest organisms. In the case of more complex animals the senses provide a large quantity of sensory impulses which, in order to be useful, require co-ordination and interpretation. If every sensory impression produced an independent response, many of which would contradict others, the result would be chaotic. Hence the need to analyse and sift the sensory input, at least to some extent.

This need was originally met by the development of small groups of nerve cells, called ganglia, with limited functions, which are connected by nerves to the appropriate sense organs, and are also provided with nerve connections with each other. In *Hydra*, a primitive kind of invertebrate, the sensory cells are connected with each other in a kind of network which covers the whole body under the surface layers.

In *Planaria*, a flatworm, this network has been simplified into a double column of nerve fibres which connect the various ganglia in which the sensory nerves terminate. These ganglia form a double row, connected side by side with transverse nerves, like a ladder—the main fibres being the uprights and the cross fibres the steps.

This is the general plan of the nervous systems of invertebrates, modified in some cases by the occurrence of specially large ganglia which deal with sensory messages from the more highly developed sense organs. For example, the front end of *Planaria* has primitive 'eyes', which are capable of distinguishing light and dark and to some extent the direction of the light and these have specially large ganglia which are connected with the ladder system, which provides contact with muscles. This enables the animal to move away from the light.

In all these organisms the nervous system is spread throughout the organism and is incapable of developing into a real brain. This began in the vertebrates in which the nerve cells and their connections were concentrated in a single column, the spinal cord, which was provided with a protective covering of vertebrae. As senses developed so did groups of cells in the spinal cord, especially at the anterior (head) end—to deal with the sensory input and to process it. The final result of this fantastic development, the brain of man, contains many billions of nerve cells (neurones), connected with each other by almost innumerable nerve fibres; some nerve cells may be connected in this way with thousands of other cells. The number of interconnections is thus almost inconceivably large.

In the higher animals the brain is one of the largest organs of the body which requires one-third of the whole blood supply (and the oxygen which it carries). It is a complex organ of which different parts have distinct functions and something has been learnt about various activities of the brains of humans and of the higher animals but it is probable that an adequate account of what they can do and how they do it lies far in the future. In a book like this, one can only attempt to mention a few of the more significant abilities and some attempts which have been made to explain them.

10 Interpreting sensations and ordering actions

We have seen that life has developed many extraordinarily efficient mechanisms for living, for obtaining food and using it and for propagating itself. How these mechanisms work is being explained more and more in terms of the behaviour of the molecules of which the living organisms are constructed and we must ask whether this is the whole story. Are animals, including human beings, simply to be regarded as physico-chemical automatons or robots, machines of almost infinite complexity, but still machines, or is there some room for still regarding human beings as responsible creatures which can direct or guide their bodily machines to do what they want to do?

In order to help us form an opinion on this question we shall first consider how actions are organised in animals of various degrees of complexity and this involves knowing how they organise their knowledge as well, since most actions take place on the basis of knowledge of one kind or another.

Broadly three types of response to a given situation are found: firstly, the *instinctive,* in which an animal responds with a definite action to a situation it may never have experienced before; secondly, the *learned* response, in which it makes use in one form or another of what it has learnt from the results of previous attempts to cope with a similar situation; thirdly, responses which are partly instinctive and partly learned.

The instinctive method of controlling actions reached its greatest development in insects, which carry out extraordinarily complex actions in their life cycles without 'learning' how to do them and obviously without having any 'understanding' of their purpose. J.H. Fabre and other students of insect life have recorded many remarkable examples of instinctive actions. For example, the hunting wasps, *Ammophila,* lay their eggs on the top of a living caterpillar, which they have paralysed by injecting a nerve poison precisely into the nerve centre of each of the caterpillar's segments. The caterpillar is then placed in a hole in the ground, which had been previously prepared and the egg layed on it. A supply of fresh food is thus provided for the larva when it hatches out.

Another wasp, *Eumenes,* constructs a clay urn, the lower part of which is filled with live caterpillars. Suspended from the top by a thread, out of reach of the caterpillars, is a single egg. When hatched, the larva hangs by the thread and is able to reach the nearer caterpillars without danger to itself. When it is bigger and stronger, it can leave its thread and eat the remainder.

Fig. 22   The wasp *Eumenes coarctatus* provisioning
her nest with a cater-pillar. (From *Wasps*
by J.P. Spradberg.)

In these and many other similar instances it is obvious that the insect performs a quite extraordinary series of actions without any knowledge of the purpose of what it does. It has been shown that each series of actions must come ın a definite order and a new stage cannot be begun until the last is completed. If the insect is prevented from doing one step in the sequence, it is unable to go on the next. Every completed step provides a necessary stimulus for the next. If the clay pot is broken the wasp may mend it, but is unable then to place the egg inside.

Nevertheless there are many examples in which the instinctive action is modified by the circumstances.

The digger wasp, *A. pubescens,* studied by G.P. Baerends* lays her eggs in several holes in the ground, which she provisions with caterpillars. She may have three different nests at varying stages at the same time and she visits them all at intervals. On some of these visits she brings additional caterpillars or moth larvae, on others she merely examines the nest to see if an egg has hatched. If the experimenter removes the food supply from a nest, the wasp promptly replenishes it. Her behaviour is thus determined to some extent by her estimate of the situation.

An Indian mason wasp, *Eumenes conica,* studied by R.W.G. Hingston†, builds clusters of clay cells, deposits an egg in each and fills it with caterpillars before sealing it with a lid. If a cell is broken, and a caterpillar falls out, it may be unnoticed, but if another gets stuck in the hole, with part of its bodẏ hanging out, the wasp attempts to push it back and then goes away to collect a pellet of clay, which it brings back to mend the hole. Another wasp *Ryrehium nitidulum* also builds pots which are

*J.P. Spradbery. *Wasps.*

†H.E. Evans and M.J.N. Emerhard. *The Wasps.*

covered and sealed with resin after having been filled with caterpillars. If a hole is made in a pot, the wasp attempts to mend it with bits of resin and finally with clay. These are all instances in which the instinctive action is modified by the local circumstances.

In the higher animals, including man, although learned responses have largely taken over as the method of adult life, some instinctive actions remain, which are performed without a period of trial and error, and without being learnt. The necessary nerve connections must be present at birth waiting for an appropriate stimulus to activate them. Thus newly born babies like calves, foals and rabbits are soon able to suck from their mothers, the feel of the breast is sufficient to activate the muscles used for sucking. Young birds will gape i.e. open their mouths for food whenever a dark object approaches the nest. Chickens will peck at any small bright object soon after being hatched. The red-backed shrike has an innate tendency to impale its food on a sharp thorn; nearly every kind of bird constructs a nest having a characteristic shape and structure. In all these cases the patterns of action must all be present in the brain, 'imprinted' in nerve connections which have been inherited and only require the appropriate stimulus to bring them into action—but nevertheless capable of being modified to some extent according to the circumstances. A foal may be able to use its legs soon after being born, but where it goes will depend on the circumstances.

Students of instinct, such as Nicolaas Tinbergen, Konrad Z. Lorenz and Bierens de Haan have demonstrated that to initiate an instinctive action a particular sensory stimulus (the releaser) is required which releases the series of actions which follow. The function of the releaser is to set in action the series of nerve impulses which cause the required muscle contractions. This is probably effected by removing some inhibition which prevents the functioning of the nerve fibres which are involved. In these cases the nerve fibres are capable of functioning without a period of learning by trial and error when the inhibition is removed. Nevertheless it has been shown that in some cases an element of 'learning' occurs because the instinctive action has to be fitted into the local circumstances and this may still involve some degree of trial and error.

We do not know how this comes about but there is no more difficulty in supposing that genes control the nerve patterns than that they control the patterns on the wings of moths. We must suppose that inheritable variations of the nerve pattern occurred during the course of evolution leading to the survival of superior patterns which resulted in advantageous types of behaviour.

While instinctive behaviour has the great advantage that no period of learning or only a short one, is required, it naturally results in an inability to cope with changed circumstances. In the long run a different method of controlling actions—in which actions are initiated in the light of the information the animal has been able to obtain about its surroundings and also in the light of its experience of the results of similar actions in the past proved to be enormously superior to the instinctive process because behaviour could be modified as circumstances change and this eventually led to intelligent behaviour.

In contrast to instinctive actions, learned responses are not spontaneous and automatic. If you put an animal which is capable of learning in a situation it has not been in before it does not immediately respond by making an interpretation of the situation, which leads to a coherent action. In fact interpreting sense impressions has to be learnt and the interpretation which is made depends on what has been learnt about previous experiences. Learning how to interpret its sense impressions is the first necessary task of the young animal and by and large it goes on throughout life. Thus, the most important task of the human child during the first few months of its life is undoubtedly to learn to interpret its sensations. It gazes intently at bright objects and tries to touch and feel them and if possible to suck them. It is trying to associate its visual data with other sensory impressions. At the same time it is trying to perform actions such as touching the objects it sees and grasping them so that it is at the same time learning to use its muscles so as to achieve a definite objective. Out of this activity it gradually builds up a picture of the world it is in, as an overall interpretation of the sensory input and any new sensations are interpreted in terms of this picture. The fact that the interpretation is a construct has been shown by the experience of persons who have had their sight restored after being blind from birth. At first they cannot make much sense of what they see and only gradually do they acquire the ability to 'see' objects i.e. to interpret the sense data in a meaningful way.

The ability to use muscles to bring about actions also has to be learnt. The movements of a young mammal are at first disorganised and it only gradually learns how to use its limbs effectively. It will try to repeat actions which produce satisfying results and after many repetitions they become easy. This means that the nerve connections which enable actions to be carried out are not initially present or at least not readily available. They only become established by being used but when established the action can be repeated easily. When a young kitten has 'learnt' to run, all the muscle actions, which are very complicated, occur in the correct sequence and this means that the nerve impulses, which activate the muscles, must arrive also at the correct muscles and in the right order. The whole matrix of nerve impulses can be turned on when the sensory situation calls for the particular action represented.

The mechanisms by which the ability to perform muscular actions is learnt have been much discussed. One theory to account for it which has been suggested is that every passage of a nerve impulse down a nerve increases the ease with which subsequent impulses pass so that the more often channels are used the easier it becomes to use them. It is also possible that a similar state of affairs applies to the junctions between nerves and neurones (synapses) in that the ease with which a nervous impulse is transmitted across a synapse increases with the extent it is used. In either case the result of 'learning to perform' an action is to facilitate the passage of the necessary impulses which can then be 'turned on' when circumstances, as indicated by the sense data, require it.

The way in which effective muscular actions are related to the picture of the world, which is arrived at as an interpretation of the sense data, was suggested by a Cambridge psychologist, Dr. K.J.N. Craik who died in 1935. The suggestion was that the brain keeps trying out possible actions within the 'model' or picture of the outside world which it has constructed. It attempts to forecast the results of possible actions within the model and when it has found one which appears to give a satisfactory result the action may be allowed to occur. For example, a man who is intending to throw a baseball or a cricket ball must make a prediction of the result of his muscular actions and the prediction is made within the picture he has of his surroundings. Similarly when a cat jumps on a mouse, the action is planned within the picture which the cat has of its surroundings and because the cat has learnt effective actions in related circumstances it is able to act effectively.

The beginning of intelligence is perhaps seen in the phenomenon of the conditioned reflex, discovered by the Russian physiologist, I.P. Pavlov (1849-1906) in his experiments with dogs. This was that nerve impulses leading to action (or in some cases the stimulation of glands) could be set in train not only by relevant sensations (such as the sight of food) but by extraneous sensations only slightly connected with the subsequent action. If a hungry dog is shown food, its mouth begins to water i.e. the sensations of sight and smell of food causes the salivary gland to secrete—an involuntary response which is regarded as a simple reflex result for the sensations. Pavlov found that if a bell is rung when food is produced the animal will after a time learn to associate the sound of the bell with food and will begin to salivate at the sound of the bell even if no food is provided. It has thus learnt to associate a sensation which by itself has nothing to do with food with the latter and responds to it involuntarily in the same way. Pavlov called this a 'conditioned reflex' and he and his followers regarded it as the basis of most animal behaviour.

Many similar experiments have been made with numerous other animals—some with very primitive species. Thus, in experiments on octopuses, Professor J.Z. Young lowered into the aquarium either a crab or a crab with a white plate attached to it. As soon as it sees the crab the octopus comes forward and seizes it but when it takes the crab with the white plate it receives an electric shock. This causes it to drop the crab and go back to its 'home'—a pile of bricks in a corner of the aquarium. It learns very quickly to associate this unpleasant experience with the white plate and will not then attempt to take a crab with a white plate attached to it.

Conversely animals will tend to repeat acts which produce pleasurable results. This is the whole basis of training performing animals. A trained dolphin goes through a complicated routine because it knows it will be rewarded with a fish. The repetition of reward reinforces the conditioned reflex until it becomes almost automatic.

Dr. J.B. Watson and more recently Dr. B.F. Skinner have built up the conditioned reflex as the basis of most animal and even human behaviour. All actions are regarded as reflexes of a more or less automatic character to situations in which we have been previously conditioned to expect to provide satisfaction of some kind. Thus began the great and still continuing controversy between the 'behaviourists' and those who believe that animals and human beings have still a certain amount of choice in their actions—a choice between alternative actions which they make as well as they can on the basis of the information they have about the situation.

The behaviourist denies that human beings have any real freedom of choice—their actions are determined by the circumstances and the responses which they have been conditioned to make.

I shall return in Chapter 12 to discuss the scientific basis of free will in animals and human beings. For the present it is sufficient to note that the 'conditioned reflex' is itself a complicated phenomenon and many of the profound abilities of the brain are involved in it. It obviously requires the existence of memory of past experiences and the ability to learn from experience. It must also be realised that in the experiments a single and simple situation is isolated from the whole life of the animal. In its normal life besides the experimental stimulus it will be receiving a great many other sensations and responding to them to some extent. The fact that when presented with a simple stimulus it responds to it in a stereotyped way, does not mean that is has no intelligence. The octopus may in fact still be using its intelligence i.e. its action is based on an appreciation of the situation which it has learnt from previous experiences.

11 Perception and consciousness

Perhaps a better understanding of how actions are initiated and controlled could be reached if we knew more about what actually goes on in the brain. In particular it would be useful if we knew how present sensations are integrated with recollections of past experiences and how the latter become available for use. The problem is a formidable one. In the human it is estimated that about 3 million nerve fibres reach the brain from the sense organs—of which more than a million originate in the eyes. The nerve impulses carried by all these nerves are analysed and interpreted mainly in the cerebral cortex, in man a great convoluted membrane, containing about 3×10^{10}* neurones i.e. about 95% of all the neurones in the brain. It might seem an impossible task to discover how such a complicated system operates. However it is possible to trace the nerve impulses from the sense organs back into the brain and find their destination in the cortex. It has been found that the areas of the cortex which receive messages from the different senses are highly localised; for example, the optical messages reach an area of the cortex at the back of the brain and if the retina is stimulated by a pattern of light, a similar pattern of excitation can be detected in the visual cortex.

Many studies of the cellular architecture of the cerebral cortex have been carried out, beginning with the pioneering observations of Ramon Cajal (1852-1934) at the time a Professor of Anatomy at Barcelona, who stained brain sections by the recently discovered process of Golgi, which makes visible some of the cells and the nerve fibres associated with them. For many years he studied the complicated structures which were displayed and much of our knowledge of the anatomical structure of the brain is due to him.

The general arrangement of the neurones is remarkably uniform in different parts of the cortex. It is built up of a number of layers in which the individual cells are arranged in columns. The cells are connected with each other (through synapses) by a very large number of interconnecting fibres. One neurone may have up to 60 000 synaptic contacts with other neurones—making the total number of interconnections in the whole of the cortex a quite staggering number*.

*i.e. 30 000 millions

*of the order of 10^{14} or 10^{15}.

79

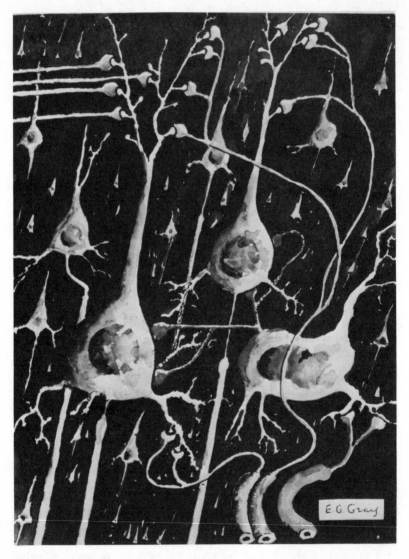

Fig. 23   A small section of the brain built up from various photographs
          to show brain cells (neurones) and some of the fibres connecting
          them, (By Professor E.G. Gray, with whose permission it is
          reproduced.)

When a nervous impulse arrives at a neurone in the lowest layer of the
cortex, it is transmitted upwards to the higher neurones in the same
column and there are also lateral contacts to cells in adjacent columns. It is

possible to study the pattern of excitation in individual cells which is produced by stimulating a sense organ in different ways, by inserting microelectrodes (very fine tubes connected with electrical recording equipment) into the cortex. Drs. D.H. Hubel and T.N. Wiesel of Harvard University have studied the patterns of excitation produced in individual cells of the visual cortex of the cat by illuminated patterns of different shapes. They discovered that certain cells (or groups of cells) are stimulated by specific patterns of light falling on the retina. Thus vertical, horizontal or oblique illuminated lines falling on the retina stimulate different neurones of the visual cortex. A similar state of affairs is indicated by the observations of Professor J.Z. Young and his colleagues with the octopus. The cells which respond to particular types of stimulus are called classificatory cells and it is likely that such cells exist which respond to many different patterns of stimulus.

We get a hint here of the reason for the enormous numbers of cells and their connections in the cortex, since if special classificatory cells or groups of cells are required to distinguish every kind of sensory pattern which is likely to be encountered, a very large number of them will be needed.

This is no doubt the beginning of an understanding of the way in which the brain classifies its sensations, but clearly neurophysiologists have a long way to go before they can understand how the visual picture as a whole is built up from large numbers of interacting elements.

About a million bits of information are being transmitted by the human eye to the brain at any one time and these may be changing very rapidly, at least several times a second. In order to reduce this input to manageable proportions, the cortex must discard much of the detail, retaining and interpreting the general outline or those parts which are of immediate interest. As a rule only this outline appears in the consciousness.

A half-tone illustration in a newspaper is made up of small dots and the differences between light and dark areas are produced by varying the number of dots in a given area. What we perceive is the overall pattern and not the individual dots. Yet the dots are easily visible if we look carefully and each one makes its separate impression on the retina. The visual cortex obviously blurs out the details of the picture and interprets the broad pattern. When an interpretation has been found the local details are unimportant.

It appears then that as the message is analysed in succeeding layers of cortical cells—the local features become of less and less important and the overall pattern becomes paramount. A square object is still recognised as square whatever the angle at which it is presented and whatever its size.

At this stage the interpretation is also compared with and fitted into the memory record of previous experiences and it finally appears in the consciousness as an overall picture of the total input from the senses. It cannot be said that science has reached or is even approaching an understanding of how this occurs or of what is involved. Indeed many physiologists deny that consciousness can be a subject for scientific investigation at all, because it is purely subjective. We have no direct knowledge of any consciousness but our own—we merely infer from what other people tell us or from their actions that their consciousnesses are similar to our own. The behaviourists say that only the acts which emerge from the sensory input are of any importance or are worthy of study. To many physiologists, the consciousness is merely a pseudo-phenomenon. To them the reality is the nerve impulses in the brain, which are real observable physical phenomena. They may present themselves to us in the form of a conscious perception but this is not what they are, but merely what they seem to be.

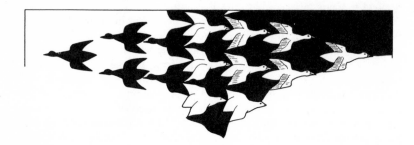

Fig. 24   A picture which can be seen in two alternative ways, either as a flight of white geese flying to the right or as a flight of black geese flying to the left. (Adapted from a drawing by M.C. Esher. *The Graphic Work of M.C. Esher.* Reproduced with permission of Ballantine Books, New York.)

On the other hand we can hardly deny that we ourselves have consciousness and it would be difficult to find an explanation of other people's behaviour except on the basis that they too are conscious of what they perceive. On the same basis we must admit that the higher animals possess some form of consciousness though probably of a more rudimentary kind than man's. Consciousness in fact seems to be the final way in which perceptions are organised and the whole of the sensory input is reduced to a single 'picture' and so made manageable. If we accept this we must admit that life has developed in consciousness a most effective means of presenting the sum total of the sensory input (or those parts of it which are extracted as being of significance) in a usable form—perhaps the most momentous development of life in the whole of evolution. The basis of this mode of perception is completely unknown, and it may be many years before science is able to approach the frontier between the physical events in the brain and the consequent perceptions. We see these processes at work in the well-known drawings, e.g. the outline cube, Fig. 25 for which two equally likely interpretations are possible which we are aware of perceiving sometimes one way and sometimes the other.

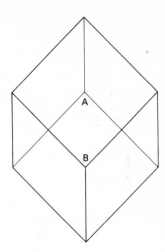

Fig. 25 Outline cube. Two interpretations are possible, according to whether corners A or B are regarded as nearer the eye.

Perhaps a starting point in correlating physical events with sensations has been found in the observations in the course of brain surgery that electrical stimulation of certain parts of the brain by means of electrodes results in the patient becoming aware of various sensations according to the position of the electrodes. In experiments of this nature made by Dr. Wilder Penfield at Montreal Neurological Institute some patients saw coloured lights, others experienced tingling of the fingers or involuntary muscle movements.

Other stimulations caused the recollection in considerable detail of long forgotten events. Dr. Penfield cites a number of instances. For example, a patient 'heard a song played by an orchestra each time the stimulating electrode was placed on the surface of the cortex'. Another patient on stimulation, could hear a certain piece of music. 'She did remember the occasion. It was Christmas Eve in Amsterdam, where she had lived. She seemed to be in a church. It was during the war and there were Canadian soldiers there'. When the temporal cortex of a young man was stimulated he cried out 'Yes, Doctor! Yes, Doctor! Now I hear people laughing—my friends—in South Africa'.

This has led to the suggestion that memories of past life are stored in great detail and with suitable treatment can be recalled. However it might be that the recall is not completely accurate, but like a dream, where images based on recollections but combined in different ways, rise into the consciousness.

When the sensory picture has been interpreted it often leads to action of one kind or another; but frequently no action follows. Thus it appears that a decision is made to act in a particular way or not to act. In the former case the motor nerves must be stimulated to bring about the necessary muscular actions. This is done by connections between the interpreting cortex and another area of the cortex—the motor area—in which the proposed action is correlated with the visual scene and other relevant perceptions. The actual nerve impulses which go to the muscles are organised in the cerebellum, a convoluted ball of neurones at the rear of the brain and below the cerebral cortex. As pointed out above, the only actions which can be easily performed are those which have been 'learnt' i.e. the necessary pathways have been established and can be 'turned on' in a block. The number of actions which a human being can perform easily is quite limited. Thus the order from the cerebral cortex may be to walk, run, jump, pick up something, stand up, sit down or lie down, convey food or drink to the mouth, masticate, etc., all actions which require very little attention, but can be modified according to the circumstances. In human beings there is also speech, with which a whole area of the cerebral cortex is concerned, a process by which a large number of meaningful sounds are produced, related closely to the situation as it is appreciated and even more to the recollected background of previous experiences. The nature of language is discussed at greater length in Chapter 13.

The behaviourist view is that speech is an action, which like other actions is an automatic response to a situation, which is made because it is capable of producing results which we have been conditioned to think of as rewarding. But frequently no speech results, perhaps because there is no-one who can hear it. In this case it remains as a thought, a tryout of a speech, a formulation of a possible statement which need not be uttered. But even when there is an audience, the speech (or other action) may not be made because some thought occurs which inhibits it. The point is that to speak or not to speak, to act or not to act, depends on a very complex situation, including recollections of past actions, some possibly quite remote in time. There appears to be a faculty which reviews the proposed utterance and decides whether to make it or not. It could be that out of the large amount of experience stored in the brain, some of which can be used in making a decision, some thought arises which inhibits further action. The common sense view is that a decision has been made—the behaviourist say that the outcome was decided by the brain as a necessary consequence of the present sensations and the store of past experiences, which is inevitable for the given situation. Further discussion on this controversy and on the possibility of free will is deferred until the next chapter.

The above is clearly a very simplified account of what may be called the 'higher' functioning of the brain. It omits many other features which are important for maintaining the life of the organism e.g. those regions which give rise to 'homeostasis' i.e. preserve the internal milieu of the organism within a range in which life is possible; those which control more or less automatically the functioning of internal organs and which give the body some knowledge of its own state. It also omits any account of 'feelings' and 'urges,' possibly relics of a more primitive type of brain which did not allow a thorough analysis and interpretation of the sensory input, in terms of recollected experience.

I have also not attempted to explain in what way the 'memory' record is made and kept. While there are numerous hypotheses about this, none has convincing support and it remains almost totally obscure—a fact which serves perhaps better than any other to demonstrate our great ignorance of the basic modes of operation of the brain.

Consciousness

Some further comments on the nature of consciousness may be added. It is the most difficult problem in the whole range of science and may be insoluble because there is no way of measuring a perception—or indeed of displaying what is perceived, to any person other than the participant. If you look, for example, at a green object you are aware of a particular sensation which you refer to as seeing something green. You know scientifically what causes this sensation e.g. the impingment of light of certain wavelengths on the retina and the consequent events in the nerves and brain are known to some extent, but the actual sensation cannot be brought into the scientific picture at all—it cannot be described in any physical terms; it cannot be measured (although it can be compared with other sensations); it cannot be described in terms of atoms or of radiations. It thus belongs to a different category from the scientific world of atoms, molecules and radiations, yet one must admit that it is intimately connected with certain conformation of nerve impulses and the consequent states of the brain.

So we find ourselves in a dilemma. Must the scientific picture of the world stop short of accounting for what we know best, indeed our only direct knowledge, namely our perceptions and feelings. Must we accept a dual world, an outer world which is the subject of science and an inner world which although connected with it in some way is incapable of being described in the same terms? The time may come when we know the physical events in brain cells which accompany the sensations. But this will not explain the nature of the sensation, only its physical background. However, we shall have to admit that certain events in brain cells are intimately associated with sensations, and this suggests that they may only be two different aspects of one phenomenon.

We should then conclude that perceptions and consciousness in general are phenomena which are associated with certain states of the atoms and molecules of the brain.

If we look back down the evolutionary scale we cannot easily detect a point at which consciousness begins. It is difficult to deny consciousness to the higher animals, yet most people would deny the existence of consciousness in plants and other primitive forms of life. In fact, it would appear that it occurs only in forms of life which possess a need to co-ordinate the complex messages received from sense organs.

The sensation is obviously intimately associated with the physical apparatus and may be regarded as a consequence of its stimulation in certain ways. The ability to produce such a sensation must be present in the physical apparatus of the brain, and it must use an ability which is present somehow in the atoms and molecules of which brains are composed. We are obliged to conclude that even atoms and molecules when arranged in certain ways are capable of experiencing sensation.

We cannot say what is the simplest arrangement in which this occurs. We cannot detect any signs of feeling or consciousness in a pound of salt. If they exist they are of a most rudimentary description. Only when we study the behaviour of animals with brains, do we find any clear evidence of perception and consciousness, yet it must be based on a property which is inherent in matter from the start.

So we come to the conclusion that the ability to perceive and feel is a fundamental ability of matter, which in the lower animals possibly only provides just glimmerings of sensation, and which only reaches a significant level in the higher animals. This capability must be a basic property of the universe, a potentiality which existed from the start and is inherent in its nature. At present we are incapable of describing it in any of the terms used for describing material bodies.

The old dualist view of life regarded men, at least, as made up of two entirely different kinds of thing, 'mind' and 'matter' which were able to influence each other in unknown ways. The progress of science, as we have seen, has steadily pushed back the idea of a special vital principle in living things, as more and more extensive physico-chemical explanations of vital processes were arrived at until most scientists feel now that there is no longer any need for it. Yet as I have tried to show, there is something which cannot be described in mechanical terms, and at present is totally separate from the picture we make of the physical universe, and this has as much right to be regarded as a fundamental phenomenon as does matter.

Instead of rejecting 'consciousness' as a fantasy or chimera we must admit its right to a real existence and we can find no point at which we can say, here sensation and feeling begin. Thus regarding matter and consciousness as two aspects of one phenomenon we should not regard either aspect as secondary and explainable in terms of the other. What the phenomenon is, and how the two aspects are connected with each other, are still a profound mystery. The time may come when we can make a little progress in connecting matter and consciousness, but that time has hardly arrived yet. To sum up, we must regard awareness and sensation and the whole world of consciousness—not as a chimera, a mere illusion, but something which has a real existence. We cannot say where mind and sensation begin because in a sense they are everywhere*.

*This theory, known as panpsychism, has been advocated by a number of authors. For further discussions see F.J. Ayala and T. Dobzhawski (eds.), *Studies in the Philosophy of Biology*, and B. Rensch *Biophilosophy*.

12 Machines and organisms

We have found in living organisms many examples of chemical mechanisms of a most intricate kind and no doubt there are many others, which remain to be discovered. Biochemists and biophysicists endeavour in their investigations to isolate from living organisms the various 'systems' which perform definite functions. Biochemists in particular attempt to make isolated systems which will perform the same function in a test-tube as they do in the organism and they tend to regard the whole organism as the sum of the isolated functions. Biophysicists may try to isolate and examine a single process in the organism, as for example when they study the processes going on in a small part of the brain or in a piece of nerve fibre.

This approach has had a considerable measure of success but there is no doubt that it falls short of complete success, except in dealing with the very simplest forms of life. The Russian biophysicist, Academician V.A. Engelhardt has pointed out* that the behaviour of any functional system depends on its interactions with the numerous molecules which surround it in the living state and these represent and are responsible for a certain range of possible functions. When two such 'systems' are present together the range of possibilities is greatly increased. Not only is each capable by itself of a certain number of interactions with surrounding molecules, but they may be capable of interacting with each other and so modifying each other's behaviour. In the whole of a living organism, even a simple one, there are a great many separate systems so that there is an enormous number of possible interactions, besides those which occur in the isolated systems. These interactions are in fact essential for the continuance of the living organism—yet the great success of biophysics and biochemistry in interpreting life on the basis of their studies of isolated systems shows that in many fields of study the behaviour of isolated systems is of paramount importance and is not completely swamped by their mutual interactions. Yet the total explanation of the functioning of the whole organism often remains elusive.

Does this mean that we must accept that organisms do indeed involve processes which are not explicable in terms of physics and chemistry? I do not think so. As time goes on we may hope for increasingly comprehensive accounts of living processes. Although the living state is one of enormous

*European Biophysics Congress, Baden, 1971.

complexity, the forces and structures are still those determined by physics and chemistry.

We therefore do not find it necessary to invoke a 'vital principle' in the sense that this term was used in the eighteenth and nineteenth centuries. The true 'vital principle' is the establishment during the course of evolution from the unknown earliest beginnings of the molecular patterns and organisations which are capable of guiding the formation of organisms which can function efficiently in their environment and are also capable of reproduction.

To account for the working of such organisms, we do not need to invoke forces other than those already recognised in chemistry and physics. Yet these forces, working through the long ages of evolution, have given rise to complex types of behaviour which could hardly have been predicted from a knowledge of the simple systems. Complex structures and types of behaviour have thus emerged from the mutual interactions of macromolecules in complex systems guided only by the Darwinian principle that only sufficiently efficient organisms survive, while non-efficient organisms are eliminated. We can recognise as the vital principle the presence of molecular patterns which were capable of continuing and of perpetuating themselves and also of accommodating the new potentialities of living which emerged from time to time.

It is thus likely that the raw material of evolution—the processes which emerged during the course of evolution—are based on forms of organisation which were inherent in living matter from the start. Probably all the basic processes of life e.g. the ability to make proteins with special properties based on the use of a nucleic acid code, were present in living systems from a very early stage and certainly in primitive unicellular organisms, long before any complex organisms appeared. Nucleic acids perhaps represent, because of their shape, a natural and inevitable mode of aggregation of the building blocks (nucleotides) and the same might be said of proteins (amino acids). In the same way, membranes are a natural mode of existence of fatty substances. Although the initial stages of the development of life can only be seen dimly at the present time, it seems likely that all the basic processes were elaborated during the earliest stage and that they already possessed the enormous potentialities which were later realised.

We must thus regard a living organism as an assemblage of chemical structures, which can perform the complicated chemical processes which enable the organism to maintain itself in a suitable environment and also to reproduce itself. They are the result of enormously long processes of natural evolution which has resulted in the assembly in each cell of a large number of functions which operate in harmony with each other. Even comparatively simple organisms like snails or earthworms or even the unicellular protozoa, could not possibly, even with our extensive knowledge of their structures, be constructed in the laboratory at the present time or in the foreseeable future owing to their very complex character.

There is perhaps no fundamental difficulty in creating living forms apart from the enormous magnitude of the task. It is possible to synthesise sections of DNA and the chemical synthesis of a small protein molecule (e.g. insulin) has been achieved. To make a unicellular organism we should have to create probably a few thousand proteins and their corresponding DNA's and assemble them in a functioning whole. This task could be shortened by making the DNA's and then introducing them into living cells so that the protein making machinery in the latter could be utilised to produce the proteins.

There is clearly a profound difference between living organisms and machines, such as we can envisage at present. A machine is an artefact which is constructed to perform at most a limited number of functions. A living organism is capable of carrying out all the functions required to maintain itself in its environment, including the ability to replicate itself. No machine yet constructed comes anywhere near to being capable of carrying out such a programme.

So although living organisms may be regarded as chemical machines, they are of a totally different order of complexity to that of any machine yet constructed or likely to be constructed. The talk in newspapers and films of a robot or mechanical man is as grotesque a misrepresentation of man as it would be of a worm or a snail or even a bacterium. It is fostered mainly by those who want to reduce the stature of man for the purpose of making human beings malleable and manipulatable and possibly also those who want to inflate their idea of their own importance by claiming the ability to achieve in their laboratories what 'the life-force' (as Bernard Shaw called it) has achieved in its innumerable experiments over the last three billion years.

There is also the fact that the higher animals and especially man have developed a means of making a record of their experiences and of creating a picture of the world, which they can examine and analyse and learn how to use and thus to acquire knowledge and understanding of the world and of themselves.

Is free will possible?

We can now ask whether, if animals and also human beings are in fact composed of chemical 'machines' in which every part behaves according to its physico-chemical nature and according to the various influences acting on it, they really have any power of influencing actions or if such actions are not inevitable consequences of the sum total of the influences acting on them. Do they have any free will in making a choice between the various possible actions?

The apparent ability to choose varies greatly in different kinds of organisms. Some actions are almost automatic responses to a sensory stimulus, others are influenced to a greater or less extent by learning from previous experiences. In such cases the action is not a direct response to the sensation but is modified by the memory records of earlier experiences, so that an action is the result not only of present sensations but also of what has happened in the past. The conditioned reflex mentioned above is a case in which an action is brought about mainly by the recollection of a previous association.

As I have already mentioned, behaviourists, led by J.B. Watson and more recently by B.F. Skinner regard all responses of animals to their environment as either direct and involuntary reflexes or conditioned reflexes which are triggered off by associations which have been learnt. This is perhaps a useful way of looking at comparatively simple situations but it is not very useful in connection with the extremely complex sources of behaviour of a human being for whom every action involves to a greater or lesser extent his whole life and experiences.

According to the behaviourists, actions are determined by the sum total of the conditioned reflexes which have been acquired by the animal or human being, each of which could be considered in isolation from the others. This is a rather simple view of mentality as there is no doubt that different interpretations interact with each other, so that most actions are not the result of a single reflex, but emerge from a complex of many elements of learning and interpretation, which together form the mentality of the animal or man.

The relevant sensations enter into an area in which usually many others are also being received and analysed. They are being interpreted in the light of past experiences, of memories which may arise momentarily into the consciousness or of set attitudes which have become imprinted on the brain. As a result of all this activity certain actions may emerge or alternatively no action at all may occur.

Under these conditions it is evident that our view of the process is very different for example from that of a machine analogue in which a sensory signal activates a circuit or a number of circuits or in which a number of signals have to be received to bring about a physical activity.

The action which emerges from a human being is determined not only by the immediate sensations but by his past experience, how he has organised them and what he has learnt from them, and what he foresees to be the consequences of any action he might take.

The action which emerges is thus a product of the individual's whole experience and ability to interpret it, i.e. what we may call his personality. The life of an animal or man is not made up of isolated acts, it is a continuous activity in which past and present are always blended. The degree to which actions are influenced by learning from the past of course varies enormously, but in the higher animals and especially in man whose actions are largely controlled by learning, we are justified in regarding the complex of past experiences and learnt abilities with present sensations as a unique matrix which is capable of varying responses and this provides freedom of choice to the extent that the animal is capable of distinguishing different possibilities. This is exactly what is meant by free will.

We can conclude that even if human beings are intricate assemblages of physico-chemical mechanisms, which nevertheless are capable of 'memory', i.e. of recording in some way previous experiences and of reviewing their mental records in such a way that their behaviour is influenced by them, we may accept that they are still capable of possessing a true personality and of making real decisions.

13 Language and communication

We can now leave the rest of the animal kingdom and consider how human beings organise their experiences in such a way as to be useful to them and to enable them to live together in a society. There is no doubt that the primary ability which brought this about is the ability to communicate by means of language. This is the distinguishing feature between man and the higher apes, which can only communicate to a limited extent.

As Drs. N. Chomsky and E.H. Lenneberg* have pointed out, communication by human language is a specifically human trait, which is quite different in principle from any sub-human mode of communication even among species which make sounds as a means of communication. The human child begins to use words at the age of about one year. By the age of four he has mastered and can use with some skill the main abstract and complex structures of his language and has a vocabulary of about 1 500 words, an astonishing achievement. It has been found that this ability is not closely connected with intelligence and the acquiring of language is not greatly delayed by adverse circumstances such as deaf and dumb parents. It is not a matter of putting words together in an arbitrary manner. Certain rules of construction are inherent from the start. Broadly, every statement expresses a relation and children are only able to put words together to express relations they understand.

Besides acquiring this ability, other changes occurred in the emergence of man e.g. skeletal changes especially in the pelvis which permitted man to assume an upright posture; improved musculature especially of the hand and arm which permits the more effective use of tools; reorganisation of the hand so as to bring the thumb into an 'opposable' position with respect to the fingers, which greatly increases both the force and precision of the hand grip; modification of the vocal organs which makes them capable of producing articulations and differences of pitch. The brain also underwent great development, especially of the cerebral cortex, increasing the capacity to co-ordinate information and to make the transformations necessary for speech as explained below. In addition the emergent human being probably acquired new inborn tendencies to indulge in types of play which provide an opportunity to develop muscular skills and a close correlation between sense impressions and muscular actions.

*N. Chomsky, *Language & Mind*, E.H. Lenneburg (ed.) *New Directions in the Study of Language.* M.I.T. Press (1964).

There was also an increased length of infancy during which the infant was fed and protected and not obliged to take part in adult pursuits and this provided a long period for the gradual organisation of experience and the development of skills. This is a long list of 'improvements', and we do not know how or in what order they were achieved. It seems likely that when the pre-human apes left the trees and took to living in the savannah lands of central Africa, they entered a period of intense competitive struggle in which increased intelligence and abilities had great survival value*.

The essential skill which made human life possible was speech and the use of language. Primarily speech requires the utterance of sounds and therefore the ability to perform the muscular actions which produce these sounds and equally the ability to distinguish these sounds accurately when heard. The sound becomes associated with the object or act which it represents. To use sounds in this way requires a highly developed ability to associate objects and acts with sound equivalents which are often quite different in character from the actions or things represented. But primarily speech is an action and the human voice must be regarded as a tool which is controlled in a similar way to the use of other tools, i.e. we listen to the sounds we make and modify them according to what we hear. The muscles which control the voice are thus regulated through the sense of hearing, as is shown by the fact that deaf persons cannot acquire speech.

An increased ability to produce a variety of sounds must have been advantageous to early man, who found that sounds could influence actions and thus bring about desired results. They could be ingratiating or frightening or merely signals for drawing attention to something.

The voice thus took its place like the limbs as a means of bringing about actions and just as muscular actions are planned as a response to the sensory picture so sounds were emitted as part of the reaction to the situation conveyed by the senses. In such sound we can see the origin of language. Certain sounds became associated with certain particular situations and thus acquired a meaning so that in time a language of sounds was acquired. The significant feature is that the sound acquired the same significance for both the emitter and the hearer and thus provided a means of communication from the one to the other—a feature of the most profound significance in human life in that it permits the sharing of experiences.

As a rule an individual has no way of sharing an experience directly with others. When I see a tree I cannot convey my own sensation directly to someone else—it can only be done by translating what I see into symbolic words like *tree, tall, green, elm*. These words call up recollections of similar experiences in the minds of people who hear them. There

*See R. Ardrey, *African Genesis.*

are of course other possible means of communication—for example, I could draw a picture of the tree and people might accept the picture as a representation of a tree. But words can be put together in different ways and are capable of a very great variety of expression. In the development of language, the actual world in relation to the speaker, is replaced by a substitute world of symbols, i.e. of words which take its place. This has had the most profound consequences for human life. In the first place, it helps the discovery of the actual world, since words are themselves an analysis of the actual world. Using the word *tree,* for example, means that we distinguish a class of object which we call trees and when we qualify the word tree, we distinguish different kinds of trees, or the number of trees. Putting together words is to perform operations, not in the actual world, but in the substitute mental world of symbols and the exploration of the actual world is enormously facilitated by making use of the substitute world of symbols.

Words imply at least some sort of classification and analysis of acts and things, and this involves picking out the salient and noticeable features. The mere fact of giving a name to a thing calls attention to it, and we tend to notice things we can name and overlook those which we cannot. This is especially important in childhood, when learning the names of things is, for children, learning about the things themselves. Ernst Cassirer says, 'By learning to name things a child does not simply add a list of artificial signs to his previous knowledge . . . of objects. He learns rather to form the concepts of such objects, to come to terms with the objective world. Henceforth the child stands on firmer ground. His vague uncertain, fluctuating perceptions and his dim feelings begin to assume a new shape. They may be said to crystallise round the name as a fixed centre—a focus of thought'*.

Words in fact constitute the first science, since to name a thing is to be aware of its separate existence and to distinguish it as a member of a class of similar things. Science has always developed by the discovery and definition of names and symbols for facets of experience.

Words are useless unless they are heard and understood by somebody, so that the new world of words is not a private world, but is essentially shared with other people. In this way human knowledge came into existence as something which individuals can share, but not possess exclusively by themselves.

What is communicated is not the experience itself but the meaning it has. Suppose I see an object which I recognise as a tree. When I use the word tree it is the *meaning* which I communicate; namely my interpretation of what I see, not the actual experience.

The primary result of this communication of meaning is that the 'world pictures' of individuals are no longer distinct and private. When the

*E. Cassirer, *An essay on man.*

meaning I find in an experience can be communicated to others, my interpretation can be compared with that of others and a public knowledge is arrived at—that is, an agreed meaning, which secures the assent of a group of people.

If I say, 'There are deer on the hill,' my neighbour looks i.e. examines more intently features of his visual field to which he had previously paid no attention and says 'No, they are rocks.' I look again and try to decide if I have made a mistake or not.

Another feature of language is that it has a more permanent character than the actual world, in which objects and events are continually changing. When you use words you isolate a feature from the transitoriness of sensations, which at least is capable of being repeated and so perpetuating the experience for a time. While the *deer on the hill* are here one moment and gone the next, the words which express the experience remain in the mind and permit it to be recalled and communicated again and again.

The invention of writing, of course, enormously increased the permanence of language and created a new situation. Words also give new dimensions to the real world. The fleeting world of today takes its place among many recollections of past worlds which they recall. The past is enshrined in a timeless world of words in which all our yesterdays live on. Daphnis' flute still plays as when Chloe put it 'to her mouth and blew it as loudly as she was able and the cows heard, and knew the note of the song'. Words also permit the creation of imaginary worlds, since we can put words together so as to make statements when there are no sensations to correspond, the real world thus becoming multiple; just one example of 'possible worlds'—of which we can create as many as we wish.

The foregoing is a rather simplified view, since language is used not only to convey statements of fact (or rather the interpretations put on simple situations) but also complex ideas, with tentative qualifications, such as arise in human relations. N. Chomsky has pointed out that these cannot be accommodated by a simple scheme of equivalents between words and ideas, and for this reason machine translation from one language to another cannot be successfully performed. Moreover, language cannot be interpreted as a mere reflex result of the stimuli which are received; it is also an expression of the personality of the speaker and has a creative function which transcends the actual situation, since the latter is integrated with his total experience and outlook, before it results in speech.

This is particularly true of words which express abstract ideas—like love, strength, bravery, nobility. Such words which obviously involve the personality of the speaker help to define and clarify the concept. It becomes a useful idea—which by being shared influences human behaviour.

There is a possibility of confusion between words and things—of knowing which words stand for real acts or real things and which for imaginary ideas and abstractions. To the primitive mind, the name was often regarded as something which belonged to the thing—a kind of emanation of it. Knowing the name of an object or a person gave some kind of power over it. In the same way primitives find it hard to understand how there can be a word with nothing real to correspond to it. The Greeks were also puzzled by this problem—how, for example, the word beauty can exist unless, somewhere behind appearances, there was a real 'beauty' from which beautiful things acquired their character. This difficulty disappears when we remember that a word is not a direct representation of anything—but a symbol of a meaning which we have found in it. Thus beauty is a meaning which is found in experience, not the experience itself.

It is the ability to use language which makes human life, as distinct from animal modes of life, possible. In the first place it makes the transmission of knowledge possible because it is through language (and to a lesser extent in some other methods of communication such as pictures) that knowledge can be accumulated and shared. The result is that human beings live in a great stream of shared knowledge, most of which they are inherited from previous generations.

Individuals just participate in it for a time and perhaps contribute something, but the stream of knowledge and ideas as represented by words and concepts involving them goes on. It is impossible to conveive of any human life without this background. The basic nature of human life is not that of individuals engaged in a solitary interpretation of their experiences but of fitting their experiences into a shared matrix of words—words which they have inherited—which form the cement of human individuals into a human society. This is the only way in which human beings can live.

The final point to be mentioned about language is that it makes imagination possible and even inevitable. It is difficult for human beings not to be imaginative. It is the reality (i.e. a verifiable interpretation) which is hard to arrive at. One always has to choose between half a dozen possible interpretations. Interpreting one's experience means, in fact, judging between different possible imaginative meanings for which we have suitable symbols in our minds. Everyone knows how difficult it is to keep imagination under control and how easily it runs riot. Symbols, such as words, being themselves imaginative, lend themselves to all sorts of imaginative combinations.

In primitive societies and perhaps in the childhood of man, the distinction between the real and the imaginative is very unclear. The real world is seen in terms of an imaginative concept which may give sense and meaning to what is only dimly understood. The whole pattern of life, involving custom, myth and ritual, makes an imaginative unity from which all activities emerge.

This stream of shared knowledge has a dynamic character of its own, since although it has an existence which is independent of the individuals which are its bearers, yet its character is capable of being influenced by these individuals and at different times different components may be emphasized or suppressed. Teilhard de Chardin* called a region of shared knowledge a noosphere and since the noosphere is continually enlarging itself by drawing more and more people into its sphere of influence, he argues that as methods of communication between diverse people become more and more effective, there will be eventually a single noosphere shared by all the people in the world, which Teilhard regarded as the final outcome of evolution, the destiny of man, an ultimate approach to deity or perhaps, one might say, to the Kingdom of God.

*Teilhard de Chardin, *The phenomenon of Man.*

14 Science and ethics

The decay of systems of morality based on religion and particularly the loss of belief in sanctions such as a judgement after death and the allotment of rewards or punishments for behaviour during life, has led to a search for other possible bases of morality. That human life must be based on some rules which are generally agreed upon, is accepted by nearly everyone; the alternative is murderous chaos. If we do not accept a traditional and authoritative basis, on what principles can such rules be based? It is a fact that the general outlook of the most advanced modern societies is still based on the system of values which is inherent in the Judaic-Christian religion. Notwithstanding the attenuation of the doctrinal aspects of religion due to the growth of scientific rationality, it remains enormously potent not only in the modern world, but also in the developing countries which have only recently emerged from barbarism. It is broadly based on a belief in the supreme significance of every human individual no matter how unimportant or even apparently defective he might be. This was derived of course from the belief that life on earth is only an episode in a continuing story—a belief shared by other world religions e.g. Hinduism, Buddhism and Islam, although their practical implications are somewhat different.

It might seem to be quite easy to establish suitable rules of conduct, such as the principle that good conduct is based on giving other people the maximum amount of consideration and in return receiving a minimum amount of interference with one's own activities. This of course begs the question, as it becomes necessary to define what are the maximum and minimum acceptable in various circumstances. The framers of the American Declaration of Independence suggested that the proper principles of human relations are those which give rise to the greatest amount of human happiness and the greatest amount of individual liberty. They regarded such principles as axioms and clearly stated them as such: 'We hold these truths to be self-evident: that all men were created equal, that they were endowed by their creator with certain inalienable rights, that among these are Life, Liberty and the pursuit of Happiness'. However happiness is relative and one person's happiness may involve another person's unhappiness or may result in other persons' loss of liberty. How much unhappiness or loss of liberty in others is one entitled to cause in the pursuit of one's own objectives? It seems remarkable to us now that the framers of the Declaration did not think it was incompatible with keeping negro slaves, who were apparently not regarded as human beings.

Any principle on which rules of conduct are based must be derived from a judgment of value. Practically all societies have arrived at such principles, which are a crystallisation of their experiences over a long period and are based on their view of the nature of man—as incorporated in their religious beliefs. With the decay of such traditional beliefs and the great authority which science has acquired owing to its ability to perform wonders, we must ask if it is possible to establish scientifically a principle of *value* on which rules of conduct can be based.

Attempts have certainly been made to do so. For example, Professor C.H. Waddington* suggested as a criterion of value the principle that any kind of conduct which assists the continuance of the general evolutionary process, which has led to man, is to be regarded as good.

However, it is doubtful if any useful rules of conduct could be arrived at on this basis, because it might be very difficult to distinguish those principles which assist the general course of evolution from those which do not. Evolution has not always proceeded by a direct route and what seems to be an aberration may turn out in the long run to be valuable. It might be argued that the initial mutations which led to humanity would not have been seen to be likely to lead to a real evolutionary advance although they managed to survive.

In any case, the most human beings could do in one generation would be to preserve and assist individuals who were believed to possess some advantage and this would involve a judgement of value. Thus, we might encourage people with a high IQ to breed and help them to bring up their children, if we believed that we were assisting the evolutionary process in doing so.

If generally adopted such a principle would obviously mean killing off or at least preventing from breeding all children and adults with inheritable defects—possibly even minor defects such as poor eyesight—and certainly those with mental deficiency. There are undoubtedly people who consider this justified and who would sanction a code which permitted such measures. Others, including those who adhere to Christianity, would regard other principles such as the sanctity of all human life as an over-riding necessity. The Christian reason for this is the belief that all people, even if defective, have immortal souls. But even if this is not accepted as valid, there is the practical difficulty as to where to draw the line between the evolutionarily valuable and those which might be considered to be an evolutionary burden. Such a line would probably be elastic and capable of political manipulation and once adopted, the principle could easily lead to the elimination of classes of people deemed undesirable, e.g. as in fact happened to the Jews in Hitler's Germany, and the Kulaks, or peasant capitalists, in the U.S.S.R.

*C.H. Waddington, *Science and Ethics*.

A more modest proposition is that the object of ethical rules is to secure merely the survival of the individual and of the community or tribe. Most systems of ethical behaviour, particularly in primitive communities, come under this heading. It is entirely legitimate for both individuals and communities to accept rules which at least enable them to survive, and survival rules for the most part can be established scientifically. But mere survival cannot be regarded as the basis of human life except at its lowest and most impoverished level. A basis for a pleasurable life, or for a sufficient satisfaction in life and for scope of the exercise of at least some of the enormous range of human abilities, is required and this is beyond the scope of science because it cannot, for the most part be formulated or measured.

A somewhat similar position is taken by Professor B.F. Skinner in his book *"Beyond Freedom and Dignity"**. He says (some technical terms omitted): "Things are good or bad presumably because of the contingencies of survival under which the species evolved". This means that things were regarded as good because they had survival value and bad if the reverse; a reasonable proposition. However, Dr. Skinner does not say how "survival value" is to be assessed. Presumably the assessment could not be regarded as "scientific" except in simple obvious cases such as the observation that certain substances are poisonous and therefore must not be eaten, and the Jewish interdiction of pigmeat, probably due to its widespread infection with trichinosis or the rule, not universally accepted, that murder is bad. He accepts that judgments of good and evil are frequently communicated in speech, i.e. people are often willing to accept codified judgments of good and evil. But he goes further in suggesting that "it is part of the genetic endowment called 'human nature' to be reinforced (i.e. to find reward) in particular ways by particular things", which must mean that in his view human beings acquire an instinctive knowledge of 'good' and 'bad', a proposition which it would be difficult to maintain.

It is true that, for example, a chimpanzee and a human being, which obviously have different genetic endowments, will react differently when faced with similar situations. Human beings may also react differently to a situation owing to their different genetic characters, e.g. some will be hot tempered, others cool and phlegmatic. Yet it would be unjustified to generalise this to imply that their responses are in the main genetically determined. In most situations with which human beings have to deal their innate responses are usually very little help in determining their actions; in fact in many cases the actions which they actually take are very different to their first 'instinctive' response. In fact 'instinct' has very little to do with a code of ethics, the object of which is often to provide a restraining influence on instinctive reactions, as Freud pointed out*.

*Jonathan Cape, London; p. 104

*S. Frend, *Civilization and its Discontents.*

Another proposal, made recently by the eminent microbiologist and Nobel prizeman, Professor Jacques Monod, in his book *L'hazard et le necessité*† is that science itself provides a system of values, from which rules of conduct can be derived, which he calls the 'ethics of knowledge'. Monod draws a clear distinction between knowledge and values. He says "Knowledge in itself is exclusive of all value judgement; whereas ethics, in essence not objective, is forever barred from the field of knowledge"—a view not unlike that developed here. However, he suggests that the decision to base actions on knowledge, rather than on an intuitively held ethical system, is itself an ethical decision. It may be admitted that an action based on knowledge is more likely to succeed in its objective than one which is not based on knowledge. But he does not state how we should decide what our objectives should be, to what objectives we should give value and what courses of action we should regard as inadmissible.

Science undoubtedly provided an objective way of studying how societies behave and may be able to assess to some extent the influence of their beliefs on their way of life and on their material success. It may thus be able to give reasons based on objective study that one kind of conduct or type of behaviour is more effective than another, but this does not itself provide a system of values. We might set up as our criterion of value the principle that the types of conduct which are most efficient have the greatest value. But how shall we judge efficiency—efficient for what purpose? We might say efficient in producing goods: the best society is that which produces the greatest quantity of goods per head—a quantity which can be precisely measured. But it is well known that material prosperity does not necessarily lead to contentment. A recent example is provided by the widespread revolt of the young in the U.S.A., in a society which probably provides more goods per head than any other which has ever existed. Its most widely held belief that the main object of life is to acquire wealth in the form of goods or the means of acquiring them, is clearly not enough, since it tends to a dull uniformity, which many of the young find repulsive. We could perhaps try to quantify the question by regarding some consequences of a code of value as positive and other negative and attempt to work out an overall expression of the value of a mode of life. This would be the scientific way of assessing value, but even this would have to be based on axioms that certain qualities are valuable and others disadvantageous. Even a scientific estimation of values would then be based on prior conceptions.

How then can a code of ethics be established? Many societies have undoubtedly arrived at such a code, with varying success. They are the result of centuries of trial and error, together with the leadership of exceptional people (e.g. prophets) who may see more clearly than others the ultimate consequences of various kinds of conduct and are capable of

†translated by A.W. Rouse, Collins, 1972 as *Chance and Necessity*.

persuading their fellow men to adopt certain principles. Such principles are really based on a view of human nature which commands belief. Since numerous systems of ethics have been evolved and we cannot prove that one is better than another (although we may have our own opinions about them), it follows that there is no unique ethic. Some we may consider more effective than others but this will be a judgement based on our own set of values and our own presumptions.

It might then be said that such judgements have no real validity, because they can only be justified on the basis of values, which as I have shown, are not provable scientifically. They are in fact based on a total estimate of human life, which is partly intuitive because it goes beyond actual experience, whereas a scientific estimate is necessarily based on fragments of experience.

The scientific study of mankind is beset with very serious difficulties, which are not encountered in the physical sciences. Firstly, there is the enormous complication of human life, even in primitive societies. It is difficult, or even impossible, to isolate simple situations, which could be examined by themselves, as every aspect of life is connected with every other aspect in a pattern the origin of which could only be found in the past. The attempt to find scientific principles in such patterns of life obviously involves great difficulties. Thus, some societies practice monogamy, some polygamy. In some cases it may be possible to find a reason for preferring the one rather than the other, such as a preponderance of one sex in certain situations. But the whole tendency of anthropological study has been to discredit the idea of any universal standards of behaviour.

Secondly, as a rule experiments are not possible, so conclusions cannot be tested by experimental variation of the conditions, to see whether the response is as the theory predicts.

Thirdly, and perhaps most important, it is difficult for the investigator to be detached and impersonal in his judgements, as he would if he were studying a problem in physics. His prejudices and beliefs and his whole attitude to life is involved. Even if he tries to be impartial the matter may be prejudged by assumptions he has always taken for granted, although they may not be applicable or reasonable in the case under investigation. He may also approach the problem with a particular theory half-formulated and looking for evidence to support it. It is also quite likely that the very act of making observations will affect the behaviour to some extent.

Science in its earlier developments, which were reviewed in the earlier chapters of this book isolated questions in which considerations of value were not involved. It *was* possible to arrive at an impersonal answer to them which was equally true for all persons and which did not involve questions of right and wrong. Science isolated questions in which considerations of value were not involved. Science indeed at first encountered opposition because some people thought that the questions it concerned itself with *were* moral questions. For example, it was thought that the question whether the earth went round the sun or the sun round the earth was a moral question because, apart from the upholding of authority, it reflected on the dignity of man if the earth was not considered to be the centre of the universe. Galileo was forced to abjure his 'scientific' views and only later when the scientific picture was so clearly seen to be superior was it accepted and taken out of the realm of morals. Similarly, the theory of the evolution of man from the lower animals was objected to on moral grounds in some quarters because it was held to be contrary to a proper view of human nature.

The confusion between science and morality is still prevalent. It ran for example through the works of George Bernard Shaw, who never hesitated to judge scientific questions as if they were moral ones. His approach to every question was intuitive. Every idea was judged by whether it fitted in with his moral concept of man as a responsible and free individual. Here is an example of his argument: 'Compulsion of everybody willy-nilly to dangerous blood-poisoning repeated for every virus discoverable by the new electron microscope, to sterilisation, to extirpation of the rivula and tonsils, to birth delivery by the Caesarean operation, to excision of several folds of the bowel and of the entire appendix, to treatment of syphilis and malaria by doses of mercury and arsenic, iodine and quinine, all of which have been advocated by leading physicians and surgeons and some of them legally enforced today, and this without protest or even mention from the loudest champions of individual liberty of thought, speech, worship, trade (especially trade), marks the rise of an abjectly credulous workship of everything calling itself Science which goes beyond any tyranny recorded of the religious creeds of Rome, Mexico or Druidic Britain.'*

It is obvious that while Shaw regarded vaccination, many surgical procedures, and the chemotherapy of diseases, as objectionable on moral grounds, he makes not the slightest attempt to ascertain whether they are or are not beneficial medically in a majority of cases. The scientific view is that these questions can be judged scientifically, i.e. without any appeal to moral questions of right or wrong, and most people will regard the enormous progress of medical science in recent decades as a justification of the impersonal attitude of medical science. Shaw apparently objected to these medical procedures on the moral ground that it is *wrong* to interfere in any way with the natural processes of human beings. This is certainly a possible moral belief, but it is not one which many human beings suffering from diseases are prepared to adopt. Most people accept the common view that interference with the organism is justified if it cures disease and ameliorates life.

A similar situation has existed in Marxist Russia in recent times, where

Everybody's Political What's What.

the only scientific views which were acceptable were those which fitted in with Marxist ideology. Marxist ideological concepts were treated as scientific axioms. If the scientific theories were apparently in conflict with them they were necessarily wrong. 'When he grasps Bolshevism,' said Lysenko, 'the reader will not be able to give his sympathy to metaphysics, and Mendelism definitely is pure, undisguised metaphysics.' That is, for the Marxist the interpretation of nature must conform with his interpretation of history and the class struggle†. At the bottom of the Marxist system is a judgement of values, and Marxism is for this reason in conflict with the spirit of science, which attempts to isolate questions of fact from values, and to judge phenomena, without being concerned with human and ethical implications.

It is now generally recognised at least in the western world that scientific questions are not moral questions. There are clearly many questions which can be discussed, and to which to some extent an impersonal answer can be given, which do not involve human prejudices, ideals, political aspirations, or considerations of right or wrong. But the fact that some questions were wrongly thought to be moral questions does not mean that there are no moral questions.

Science is necessarily analytical. It collects information. In the human sciences it compares one custom with another. It seeks origins and reasons. It does not and cannot integrate such knowledge into a system of living, because it cannot decide what is valuable and worthwhile. Anthropologists may study different types of human societies, but they do it objectively. They do not seek to judge—they necessarily try to be impartial observers. They may perhaps feel that some cultures are more satisfactory than others, but that is a personal predilection. To the scientist, all ways of living are of equal worth as objects of study.

Scientific study thus gives no conclusive guidance on moral questions. It is impossible to deduce from it a criterion of value, of what is right or wrong. This still has to be obtained by human beings from an estimate of their total condition, which includes what is scientifically established and much that is not. A single human being cannot do this for himself, because human wisdom is cumulative and everyone necessarily inherits the broad conclusions and beliefs which previous generations have arrived at as the basis of a common life.

Such beliefs are beyond the scope of science because it necessarily deals with fragments of experience, while a principle of value necessarily has to deal with the whole of human experience, in so far as it can be grasped. It is necessarily arrived at intuitively because it is a leap in the dark, an extrapolation from what is actually known; while it can only be arrived at

† Dr. Zhores Medvedev was sent to an insane asylum in 1971 for his criticisms of the bureaucratic control of science which made Lysenko possible. (*The Medvedev Papers: the plight of Soviet Science.*)

by a process of trial and error, it is more than a summation of past experiences.

It might be said that any generalisation of experience is a kind of knowledge, which is entitled to be regarded as science. But considerations of value really go beyond knowledge, they involve not only knowledge of what is, but a judgement of what is best and this is necessarily a leap into the unknown and a projection into the future. To distinguish between right and wrong is not only to study mankind as it exists, but to endeavour to establish a particular sort of relation between human beings now and in the future. It was not an accident that the Hebrew preachers, who frequently denounced the evil doings of the Israelites and called upon them to conform to an exacting code of behaviour, were called *Prophets*.

15 Science and belief

If human societies require a belief or beliefs about the nature of human beings on which their relations with each other are based, we must ask how such beliefs are arrived at and what are their necessary characteristics.

During the present century, a very large number of primitive societies has been studied by anthropologists, in some cases before they were much influenced by contact with more advanced civilisations. From these studies one remarkable fact has emerged, the significance of which has not been sufficiently appreciated, viz. the life of primitive societies is rarely simply economic, but is usually based on a complicated system of beliefs. These are theories of life, which provide a background and a rationale for the customs which carry the society along from one generation to the succeeding ones. But owing to the persistent curiosity and imagination of human beings and their need to know who they are and where they come from; these beliefs are nearly always much more than a recipe for living; they provide an explanation in the form of myths of the nature of human beings, how they are related to each other, what happens when they die, and how they fit in with the rest of nature.

It is usually not known how these systems of beliefs originated; their origins are hidden by the mists of time. There is one belief which is almost universal among primitive peoples, including some such as the Australian aborigines and some American Indian tribes, who preserved a stone age culture until they received the impact of civilisation, so that it was probably widespread in the stone ages. This is a belief in a world behind the world of appearances, in which in one form or another spirits of various kinds exist—not only the spirits of departed people, but also independent spirits which might interfere with human life for good or evil, and spirits capable of controlling the inanimate world.

Belief in a spirit world behind appearances comes easily to mankind, probably because of the nature of human perception. Things seen are not so much apprehended directly, as realised, as we have seen (Chapter 11) in a kind of construct from the actual sense data. Perceptions therefore possess something of the quality of a vision. You have only to shut your eyes and the vision disappears, only to reappear in an altered form in dreams. It was easy to believe in the existence of persons and objects which could not be seen. Moreover it seemed to make sense to attribute the extraordinarily complex and incomprehensible happenings in the physical world, like storms and earthquakes, and even the familiar routines of the sun, moon and seasons, to the activities of unseen agencies, to 'spirits' which might also interfere with human activities and therefore needed to be propitiated and supplicated to.

The major religions of the world developed out of the profound belief which primitive and not so primitive man has had almost everywhere in the existence of a spirit world behind appearances, to which human beings belong to some degree and to which they might return after death. As such beliefs became more organised they took on more and more tutelary and ethical functions and incorporated to an increasing extent the ethical standards of the society.

These characteristics have been retained by the principal world religions which have based themselves on a firm belief in a level of existence beyond the phenomenological one, in which the ethical standards are determined, for example by a system of rewards for good conduct, and punishments for bad conduct.

The advance of science which has been traced in this book, has provided scientific explanations of many natural phenomena and it is no longer possible to credit that events in the natural world are under the control of supernatural intelligences.

The ordinary everyday happenings in the world were shown to be due to the automatic operation of understandable laws. It appeared to be a reasonable inference from scientific knowledge that the power behind the universe, whatever its nature, did not interfere in day-to-day events. It was remote and unfathomable and there seemed no reason to expect that it had any direct influence on events in our world, human or inhuman. So far as human behaviour was concerned it was neutral like Thomas Hardy's Immanent Will, which

    'has woven with an absent heed
    Since life first was and ever will so weave'.

The central religious belief in the existence of 'another world' and the survival of the human spirit after bodily death has also become vulnerable. The increasing knowledge of the physiology of the human body has shown the utter dependence of all human functions on the material machine, and it has become increasingly difficult to see how there could be any survival of human personality, except in the memories of other people and in concrete and lasting achievements. The whole tendency of modern science, notwithstanding its incompleteness in some directions, is to discredit any dualist position and to deny the possibility of any residue of personality surviving the dissolution of its material substratum.

If this is so, one must ask if a purely ethical religion is possible, that is, one which does not rely on supernatural sanctions. As I have suggested earlier it is not possible to derive a code of ethics scientifically; it can only come from a belief in a principle which the society regards as of supreme importance, such as the Judaic concept of justice, compassion and righteousness which is epitomised in the Jewish Jehovah; the Moslem creed of submission to the will of Allah, the Hindu belief in reincarnation or the Christian concept of another life for all human beings and hence the supreme importance of every human life however poor and insignificant.

The Christian concept of immortality has clearly become untenable but the Christian ethic is probably more influential and widespread now than it has ever been. It is certainly potent in all the Western nations, and is at the bottom of all movements to secure the rights of man and to care for unfortunate and defective individuals. It has been enormously influential in most of the African developing communities even when they do not formally accept a Christian doctrine.

There would not appear to be any great difficulty in adopting a code of ethics based on this principle without accepting the supernatural elements of traditional Christian faith, such as a personal God and life after death. In fact such a position is widely taken, tacitly if not expressly, within Christian churches and in associated societies.

Inside the Christian churches there has been in recent years a movement initiated by Rudolf Bultman, Paul Tillich and Dietrich Bonhoeffer and popularised by the former Bishop of Woolwich, the Rt. Rev. John Robinson* to interpret doctrinal elements in Christianity very broadly, and to take the central statements of faith not literally, but as symbols of profound truths. The idea of God underwent continuous development both in the pre-Christian and in post-Christian periods, and it is not very difficult to go a little further and divest it completely of personal implications and to regard God as a repository of what we consider to be the Good—or perhaps the Ideal, in human relationships and of a proper human feeling of respect towards the immensity of the Universe and the processes which have produced us.

This is not very different from the idea of God which is tacitly held by many in Christian congregations. It makes possible, to give an extreme instance, joint services between different religions which have sometimes taken place.

However, such an abstract idea may be thought to be insufficient to produce an active emotional response with many people. The idea of God is valuable because it sums up not only the majesty and power of the Universe, but also the belief that justice and compassion are central concepts in human life and a necessary basis of human relations. It thus provides something which science by itself cannot provide. Possibly the form in which the belief is clothed is not of great importance, but images which are understandable and acceptable at every age and by every condition of humans are precisely what are required of religion.

To those who say that such a concept goes beyond what is scientifically proven or formulated with precision we may point out that the scientific picture itself involves an exercise of faith. It is an act of faith to believe

*J. Robinson. *Honest to God*, 1963.

that science can or will provide anything like a complete explanation of life and the universe, notwithstanding its success so far. The solution of one set of problems usually uncovers others, which were previously unsuspected.

Science at present is very incomplete even in directions which have been intensively pursued and some of these have been mentioned in earlier chapters. For example, while biochemistry can in many cases give a good account of isolated functions, our knowledge of the total integration of all functions to make an autonomous organism is very incomplete; although the concept of evolution provides a general account of how species developed we have at present, notwithstanding the genetic code, very little idea as to how new organs and new forms of living things originated, thus it is completely unknown how the 'codes' for the long neck and other features of a giraffe or the wings of a fly are carried. We also do not know much about how complete complex organisms develop from the initial fertilised egg cell; while elementary concepts about the functioning of the nervous system have been reached, an adequate account of how they are integrated in a functioning brain is for the probably distant future; it is not even agreed that the concept of consciousness can even be a matter for scientific enquiry.

No doubt considerable advances will be made in all these directions, but it is the faith that the incomplete concepts which he has arrived at, will be part of the larger design of the future, which carries the scientist on.

In any case even if these problems were solved they would not help human beings to find meaning in life or to define the permissible in their relationships with each other.

It is reasonable for human beings to accept as an article of faith, in whatever terms it is expressed, that the Universe is not a pointless accident, although it may be difficult on some premises to provide a logical proof. The very fact of human life may be taken to provide one. Not to exercise this faith is sheer nihilism and can only lead to utter anarchy.

To believe that life in general and human life in particular is valuable and both provides and requires criteria of value is not a less useful or justifiable concept than the faith of the scientist that mechanistic explanations will be found for many phenomena which are at present obscure. Such a belief cannot be expressed in the language of science, indeed the only language available is a humanistic one and myths and parables are as valuable to human beings in general as the laws of science are to the scientiest.

It has also frequently been pointed out that scientific explanations are themselves in a sense myths, since they are human constructs which are useful in explaining phenomena in terms of simpler concepts, but which only correspond with reality up to a point.

Science operates by breaking up experience into easily manageable fragments, but human life requires an assessment of the total situation such as is provided by religious beliefs which necessarily go beyond what is demonstrable scientifically, because they deal with the only partly apprehended whole. Their real object is human aspirations, with the human situation as a whole and finding guides for human conduct, and this can only be communicated or, at least, illustrated in the form of myths.

In the elusive field of human relations and human values we must accept that myths are useful and valuable and they do not have to be erudite, in fact to be useful they must be understandable by everybody.

Outside the Christian churches, the general ethics of Christianity have been accepted by a large number of fringe religions and are largely acceptable to those who call themselves Rationalists and Humanists, whose position is a general human benevolence, based on a rational and scientific view of life. The basic principles behind these codes do not seem to have been very clearly formulated and unfortunately the Humanists devote more effort to vigorously combatting what they regard as religious superstitions than in looking for points which they hold in common with them.

An ethical code, however strongly it appeals to thinking people, leaves the imagination untouched. Ordinary human beings need, or at least respond to, something more poetic and resonating. The problem is not an individual one, but a question of how to inspire masses of people with a pattern of thought i.e. a way of looking at human experience. Christianity did this for many centuries for the Western World. While it was a crystallisation of the main lines of thought of the ancient world, it was also something new and distinctive in its emphasis on simple human relationships and on human personality. It was this emphasis which, notwithstanding many failures and bitter strife between different sects, gave western civilisation its distinctive character. In due course its concern for the individual gave rise to the principles of individual liberty and the rights of man which are basic to western civilisation and have spread through much of the world. It provided the seed bed out of which came that tremendous outburst of European creativity in politics, literature and science, which has transformed the world and which many other civilisations and approaching civilisations are at present endeavouring to assimilate on their own terms. In its attempt to illustrate the homely truths of the Christian story to the illiterate, it was the cradle of the remarkable achievements of western architecture, painting and music over the centuries.

The erosion of the Christian doctrine has led to a widespread decay of adherence to Christian ethics as well. As a result the basis on which western societies rest has become unsubstantial. They are unsure of themselves, since they do not know what to believe or what objectives to aim at.

In the last three hundred years, science has resulted in the fragmentation of knowledge into a series of compartments, each of which can only be understood by experts. This fragmentation has introduced into every aspect of life and as a result the common ground held by people of different occupations has narrowed and society tends to break up into hostile and competitive groups. Such common ground as there is comes from the past and even this is far from firmly held.

What is needed is a new covenant on the basis of human life, which is compatible with all the scientific knowledge of the era but does not accept its supremacy or its sufficiency. While it should not reject any source of human inspiration, it has at hand as its most likely basis for western people the teachings of Jesus, the man, which are already inextricably embedded in all western thought, culture and institutions and which provide a reservoir of feeling which has accumulated around them for two thousand years. If this is rejected probably nothing is left to hold together western civilisation and its final decay will be inevitable. It is fashionable especially among the young to scoff at Christianity, as represented by the various churches, and to think that science by itself provides a sufficient basis for human life. What I have attempted to show is that science notwithstanding its successes, does not by itself provide a basis for human ethics.

The bald statement of an ethical principle, however, does little to compel adherence. A living faith must provide emotional satisfaction as well as a basis of intellectual adherence. The various religions also offer to human beings the continuing support of the respective Churches, which provide a sense of continuity and a feeling of community—of being engaged in a common endeavour. They also provide in ritual the symbolic gestures which give significance to the crucial stages of life and they provide support for the individual in everyday life as well as in misfortunes.

I have endeavoured to show that Christianity without its supernatural elements has much to offer the rationalist and the Christian churches also have much to gain by not insisting on literal adherence to doctrines which have lost credibility. Is it too much to hope that the western world can gain unity again in a Christianity which does not insist on its supernatural elements?

This would seem to be the best and perhaps the only possible basis of a unification of the Christian and rationalist positions. It would mean the churches giving up many extravagences they have accumulated over the centuries and reconsidering their hostility to people who accept their ethical views but not their dogmas.

However the Judaic-Christian tradition can now be seen to be incomplete in one respect and because of this, harmful to human life in the long run, because it encouraged the pillage of the non-human world by man. The Old Testament statement that the animal world was created for the service of mankind has resulted in a steadily increasing destruction of other species and the increase in the numbers and effectiveness of human beings now threatens the survival of many wild species and is also turning some parts of the world into deserts by destroying the plant life as well. If this continues it will not only destroy the environment which provides the material basis of human life, but it will also destroy human confidence in life itself, which will no longer be seen to be a realm of infinite variety and richness, but something quite fragile and liable to extinction.

The time is ripe for a new human ethic which deals not only with human relations but also with the relations of human beings with other life forms*. It is necessary to recognise that we are part of the whole life process and that we owe a feeling of reverence and respect to life as a whole, not merely for our own convenience and self-preservation, but as something of which we are part and without which we would never have existed. There are undoubtedly signs in the numerous wild life and environment societies that this is being increasingly recognised at the present time.

I have tried to give a picture of the life process as a whole and to show that although there is much that is mechanical the higher manifestations of life are not reducible to the status of automata by the advances of science.

Notwithstanding the great successes of science, it will I hope have become clear to readers that much remains to be learnt about the 'secrets of life'. Indeed the modern biologist must still feel as Newton did when he said, 'To myself I seem to have been only like a boy playing on the seashore and diverting myself in now and then finding a smoother pebble or a prettier shell than ordinary, whilst the great ocean of truth lay all undiscovered before me'.

*See John Passmore *Man's Responsibilities for Nature.*

Selected book list for further reading

I append a list of books, mainly non-technical, which is far from complete, but may be useful for readers who wish to go further into some of the subjects discussed.

CHAPTER 1

W.R. Shea, *Galileo's Intellectual Revolution*, Macmillan (1972).
H. Butterfield, *The Origins of Modern Science*, Bell (1957).
Sir Charles Sherrington, *Man on his Nature*, C.U.P. (1942).
V. Elkann, *Discovery of the Conservation of Energy*, Hutchinson (1974).
W.P.D. Wightman, *The Growth of Scientific Ideas*, Oliver & Boyd (1971).
A.R. Hall and Marie Boas, *From Galileo to Newton*, Collins (1963).
A. Koestler, *The Sleepwalkers*, Hutchinson (1968).

CHAPTER 2

C. Darwin *The Voyage of the Beagle, The Origin of the Species.*
F. Jacob (translated by B.E. Spillmann), *The Logic of Living Systems*, Allen Lane (1974).
G.R. Taylor, *The Science of Life: A Picture History of Biology*, Thames & Hudson (1963).
J. Huxley and H.B.D. Kettlewell, *Charles Darwin and his World*, Thames & Hudson (1965).
F.R. Jevons, *Biochemical Approach to Life*, Allen & Unwin (1964/8).
C. Singer, *Short History of Scientific Ideas, to 1900*, O.U.P. (1959).

CHAPTERS 3 and 4

A more complete account of molecular biology is given in numerous books, including the author's *The Life Process* (Allen & Unwin and Basic Books, 1970).

J.C. Kendrew, *The Thread of Life — An Introduction to Molecular Biology*, Bell (1966).
Scientific American, *The Molecular Basis of Life. An Introduction to Molecular Biology*, W.H. Freeman (1965).
R. Olby, *The Path to the Double Helix*, Macmillan (1974).
J.D. Watson, *The Double Helix*, Weidenfeld & Nicholson (1968).
J.R.S. Fincham, *Microbial and Molecular Genetics*, Eng. Univ. Press (1965).

CHAPTER 5

C.H. Andrews, *The Natural History of Viruses*, Weidenfeld & Nicholson (1967).
K.M. Smith, *Viruses*, C.U.P. (1962).
F. Jacob and E.L. Wollman, *Sexuality and Genetics of Bacteria*, Academic Press (1961).
W. Hayes, *Genetics of Bacteria and their Viruses (A Comprehensive Treatise)*, Blackwell (1969).
H. Fraenkel-Conrat, *The Chemistry and Biology of Viruses*, Academic Press (1969).
H. Fraenkel-Conrat, *Molecular Basis of Virology*, Van Nostrand-Reinhold (1968).

CHAPTER 6

J.D. Bernal and A. Synge, *The Origin of Life*, O.U.P. (1972).
J.D. Bernal, *The Origin of Life*, Weidenfeld & Nicholson (1967).
S.L. Miller and L.E. Orgel, *The Origins of Life on the Earth*, Prentice-Hall (1973).
J.A. Jukes, *Molecules and Evolution*, Colombia U.P. (1966).
C. Ponnamperuma, *Origins of Life*, Thames & Hudson (World of Science series) (1972).

CHAPTER 7

H. Harris, *Cell Fusion (Dunham Lectures)*, O.U.P. (1970).
J.B. Gurdon, *The Control of Gene Expression in Animal Development*, O.U.P. (1974).
C.R. Austin (ed.), *The Mammalian Fetus in Vitro*, Chapman & Hall (1970).
D. Kennedy (ed.), *Cellular and Organismal Biology*, *(Scientific American)*, W.H. Freeman (1974).
J. Brachet, *The Biochemistry of Development*, Pergamon (1960).
J.A.V. Butler, *Gene Control in the Living Cell*, Allen & Unwin and Basic Books (1970).
D'Arcy W. Thompson, *On Growth and Form*, C.U.P. (1942).
Max Hamburgh, *Theories of Differentiation*, Ed. Arnold (1971). (Contemporary Biology).

CHAPTERS 9 and 10

D.E. Broadbent, *Behaviour*, Methuen (1961). (University paperbacks).
N. Tinbergen, *The Study of Instinct*, O.U.P. (1951).
W.H. Thorpe, *Learning and Instinct in Animals*, Methuen (1956/1963).
N. Tinbergen, *Animal Behaviour*, Time-Life International (1966).
W.H. Thorpe, *Animal Nature and Human Nature*, Methuen (1974).
K.Z. Lorenz, *On Aggression*, Methuen (1966).
K.Z. Lorenz, *King Solomon's Ring*, Methuen and Crowell (1952).
K.Z. Lorenz, *Studies of Animal and Human Behaviour*, Methuen (1970).

J.R. Wilson (ed.), *The Mind.* Time-Life Science Library (1965).
A. Koestler, *The Ghost in the Machine,* Hutchinson (1967).
F.J. Ayala and T. Dobzhanski (eds.), *Studies in the Philosophy of Biology,* Macmillan (1974).
K. von Frisch, *Animal Architecture,* Hutchinson (1975).

CHAPTER 11

A.J.P. Kenny, H.C. Longuet-Higgins, J.R. Lucas and C.H. Waddington, *The Development of the Mind,* Edinburgh U.P. (1974).
R.L. Gregory, *The Intelligent Eye,* Weidenfeld and Nicholson (1970).
L. Stevenson, *Seven Theories of Human Nature,* O.U.P. (1974).
B. Rensch, trans. by C.A.M. Sym, *Biophilosophy,* Columbia Univ. Press (1971).

CHAPTER 12

J.B. Watson, *Behaviourism,* Norton (1924).
B.F. Skinner, *Science and Human Behaviour,* Cape (1965).
A. Koestler and J.R. Smythers (ed.), *Beyond Reductionism: New Perspectives in the Life Sciences,* Hutchinson (1969).
H. Wheeler (ed.), *Beyond the Punitive Society,* Wildwood House (1973).
B.F. Skinner, *Beyond Freedom and Dignity,* Cape (1972).
M. Chance and C.J. Jolly, *Social Groups of Monkeys, Apes and Men,* Cape (1970).

CHAPTER 13

N. Chomsky, *Language and Mind,* Harcourt Brace (1968).
E.A. Lenneberg, *New Directions in the Study of Language,* M.I.T. Press (1964).
R.A. Wilson, *The Miraculous Birth of Language,* Dent. Guild Books (1949).
R. Ardrey, *African Genesis,* Collins (1961).
J. Napier, *The Roots of Mankind,* Allen & Unwin (1971).
T. de Chardin (transl. Wall), *The Phenomenon of Man,* Collins (1965).
P. Farb, *Man's Rise to Civilization,* Secker & Warburg (1969).
T. de Chardin, *Le Milieu Divin,* Collins (1968).
D.G. Boyle, *Language and Thinking in Human Development,* Hutchinson (1971).
C.K. Ogden and I.A. Richards, *The Meaning of Meaning: A Study of the Influence of Language upon Thought and of the Science of Symbolism,* Routledge (1953).

CHAPTER 14

J. Monod, *Chance and Necessity,* Collins (1972).
J. Lewis, *Beyond Chance and Necessity,* Garnstone (1970).
Z. Medvedev, *The Medvedev Papers: The Plight of Soviet Science,* Macmillan (1971).
W.H. Thorpe, *Science, Man and Morals,* Methuen (1965).

CHAPTER 15

J. Passmore, *Man's Responsibility for Nature: Ecological Problems and Western Traditions*, Duckworth (1974).

E. Cassirer, *An Essay on Man*, Yale U.P. (1944).

R. Karsten, *The Origins of Religion*, Kegan Paul (1935).

A. Toynbee, *A Historian's Approach to Religion*, O.U.P. (1956).

K. Bliss, *The Future of Religion*, Watts (1969).

J. Robinson, *Honest to God*, S.C.M. Press (1963).

P. Tillich, *Biblical Religion and the Search for Ultimate Reality*, Nisbet (1955).

D.R. Arthur, *Survival: Man and his Environment*, Eng. Univ. Press (1969).

Subject Index

Name Index